Microelectronic Circuit Design for Energy Harvesting Systems

Maurizio Di Paolo Emilio

Microelectronic Circuit Design for Energy Harvesting Systems

Maurizio Di Paolo Emilio
Data Acquisition System
Pescara, Italy

ISBN 978-3-319-83775-8 ISBN 978-3-319-47587-5 (eBook)
DOI 10.1007/978-3-319-47587-5

Softcover reprint of the hardcover 1st edition 2016

Printed on acid-free paper

This Springer imprint is published by Springer Nature
The registered company is Springer International Publishing AG
The registered company address is: Gewerbestrasse 11, 6330 Cham, Switzerland

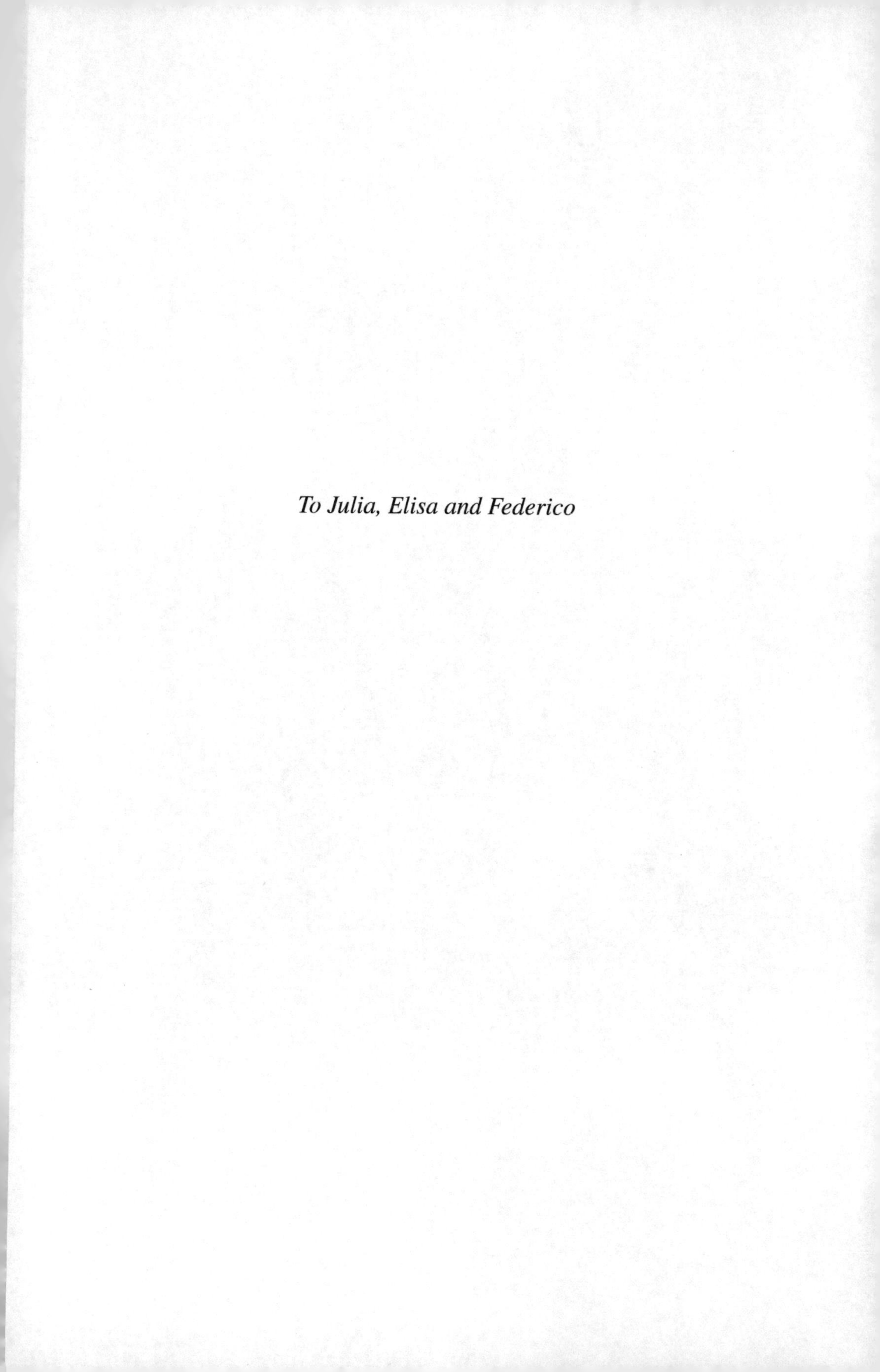

To Julia, Elisa and Federico

Imagination is more important than knowledge. [A. Einstein]

When wireless is perfectly applied, the whole earth will be converted into a huge brain, capable of response in every one of its parts. [N. Tesla]

Foreword

Energy is everywhere, sun, wind, temperature and other sources rarely known like earth vibrations: all of these are sources of "free" energy. There are not fuel costs. The only investment required is facility and maintenance. Modern technology has a lot of "waste energy" that needs to be captured. Producing "green energy" and harvesting waste energy are the solution to create a better and sustainable world! Thanks to Tesla, it is possible to gain energy at zero cost, but we don't know how to obtain it. There is only one solution to resolve this problem: research and development. Energy is everywhere, as Tesla said, from the most simple resources to capture (sun, water, wind) to more complex forms to harvest (vibrations, heat, sound, movements). Energy harvesting is the solution for small and large energy requirements: from smartphones to automotive. Thinking about the automotive market, electric cars were invented more than 20 years ago, but only now there are companies like Tesla Motors that combine technology and design. They gain billion dollars in a few days. Energy efficiency is a critical point for electric cars and a big contribution was given by the development of battery technology and their evolution. In addition, features like "start and stop" or "regenerative brake" or even energy studies about "regenerative shock absorbers" were introduced only in the last few years. The lesson that we learned is to optimise the energy recovery from a mix of sources and not from a single source. This is the way to obtain the best performance. How often does a smartphone vibrate and how many times are they illuminated by sunlight or artificial light or even are they exposed to body heat. . .all sources of energy! *Microelectronic Circuit Design for Energy-Harvesting Systems* is the ideal guide to explore and examine in detail this technology. It is only a matter of time when smartphones without battery will be produced. Understanding energy harvesting and its power management means discovering the future of power supply.

The author is the right person to discuss these topics. He is an electronic engineer and also a physicist, so who is better than him to accompany us in this study?

Emanuele Bonanni

Founder and Editor
Elettronica Open Source, EOS Book
and Firmware Magazines
Rome, Italy
September 2016

Foreword

My father was a bit of a mathematician, physicist and all-round curious sort. I grew up listening to him about scientific matters that a 6- or 7-year-old may have not immediately understood but does get interested in, enough to seek them out when older. One such concept my father spoke of and that stuck in my mind was perpetuum mobile or perpetual motion, which allows a machine to run unabated forever without running out of power. For a very long time I believed in this machine; it took a great deal of learning and understanding of the physical world to realise it is not possible eventually; everything runs out of energy for a variety of reasons. However, one type of energy does convert into another, allowing for something to act, move or operate. Maurizio's latest book discusses the concept of energy harvesting, where power may be collected from a viable source, which could be as simple as a person walking or just waving a hand. Of course, in addition to motion or vibration, other unlimited sources of energy could be light, sound or heat abundantly available from the environment around us. Imagine what a single car could power if only a fraction of its unwanted outputs are to be harnessed, i.e. collected for use by another device, most likely electronic! The impact of energy harvesting could be huge economically and, more importantly, for the environment, which we must start thinking about how to best protect and preserve. Maurizio is addressing all these issues in his new book, where he also focuses on different types of technologies available for energy harvesting. He describes the basic and advanced concepts of energy harvesting in terms of physics and engineering and proposes design techniques suited to power supplies for low-power systems. In his book, Maurizio introduces different principles of transducer operation and materials and related devices, as well as power management, storage and design of energy-harvesting systems and their future architectures. I am excited about this future one that will be so greatly enabled by technology. I am certain that 1 day an electronic device or system will be able to swap seamlessly between energy sources for its power, so it can run forever a little like perpetuum mobile. Who says this is not doable? I can hardly wait to see what happens next.

Editor, Electronics World
London, UK
September 2016

Svetlana Josifovska

Preface

Today's technology world is certainly evolving and is also bringing most innovative devices that have ever come to the market. One of those is called energy harvesting. Energy harvesting is the process of taking energy from external sources and converting it to electrical energy to supply any mobile device. Michael Faraday's law of induction found that moving a magnet though a loop of wire would create an electrical energy in terms of current. This principle can be one point of starting for energy harvesting. In the early twentieth century, a great scientist tried to pose the question to which we seek an answer yet. We all live in a solar system with an enormous amount of energy sent to the earth; adequate protection system (see the atmosphere) allows us to live and prevent the destruction of the planet under the powerful yield of energy from the sun. In addition to this, the earth is a living organism, or an accumulator and generator for various types of known and unknown energies. The question is legitimate: is it possible to be able to "catch" this enormous energy and make it available to users? Scientifically speaking, yes, it is possible; the thing is possible. A classic example is the photovoltaic cells that convert solar energy into an electrical signal (photovoltaic effect). This question was asked by Nikolas Tesla. Currently, most of these electronic devices are powered by batteries. However, batteries have several disadvantages: they need to be replaced or recharge periodically and mostly they are not handy with their size and weight compared to a highly electronic technology. One possibility to overcome these power limitations is to extract energy from the environment to recharge a battery or even to directly power the electronic device. The environmental energy is naturally occurring in large and micro-scale; the technologies have been widely disseminated efficiently: solar energy is an example that although the overall efficiency remains remarkably low (around 30 %), its usefulness is much appreciated, or almost. Fossil fuels are limited, expensive and, above all, not environmentally friendly as they induce a strong impact on land-based pollution. The photovoltaic system is a classical green system that converts solar energy to electric current to supply electronic devices. It needs improvements in terms of efficiency and new materials with the goal to be a system totally dependent in periods of minimum intensity of the sun. In this context, however, it is fundamental that the battery management system have a great capacity and excellent long-term efficiency. How much energy is available around us? What kind of energy sources do we have? What is utility? They are some issues to which we will answer in this book through an engineering discussion with the basic and advanced concept about physics and electronic circuit. In addition to large-scale energy such as solar, there are variants of energy, which could be defined on a small scale to implement in low-power systems such as wearable and smartphone devices. Walking can also be used to produce energy by using an electromagnetic mechanism. The electronics and microelectronics are spreading steadily, and many companies provide day after day IC systems of energy harvesting for different types of energies such as electromagnetic. The purpose is always to harvest the energy dispersed in the environment for reusing it in other forms (electric current) to power other electronic devices or the same device in a way that we could define recycling energy loop. Collecting all these energies (heat, light, sound, vibration, movement) could have a significant impact concerning the economic and environmental factors, reducing costs and developing new sensor technologies. The main part of every electronic system is the battery, as in a computer or a smartphone, and thinking to recharge it by external source of energy in a harvesting automatic mode could be very impressive with zero-impact work process: a smartphone supplied by environmental energy without battery. Eliminating the battery is a long-term goal that in some systems such as photovoltaics is definitely an essential element in the design. The physical aspects that come into play in an energy-harvesting system can be described in terms of ability to store the energy, materials science, microelectronics for power management and systems engineering. All electronic systems such as computer and smartphones waste energy: why not charge your phone by using its electromagnetic waves that we know to be of greater intensity during calls and receiving data? Still, why not detect the energy that the universe sends to us, such as cosmic rays for the realisation of a low-power system to supply wearable systems. But it is interesting to note that there are other sources that have emerged from the action of man, as a consequence of industrial and technological development. These modern energy sources (or artificial) are directly related to energy harvesting; vibration or temperature gradients are produced by machines and engines. Even in the electromagnetic spectrum, we can collect the energy not only from the natural solar radiation but also by all the artificial radio sources which is acquiring a great importance with the development of web-based devices concerning IoT and IIoT. The technology behind energy harvesting is possible, thanks to a careful analysis and design of power management factors that have reduced the consumption of electronic systems. Although manufacturers struggle to reduce battery consumption, running out of power after just a few hours of use and having to be connected to a power supply to recharge are common problems that need a solution. The goal of the book is to focus on energy harvesting which is released into the environment in various types: electromagnetic, vibrational and heat. The most used sources are vibration, movement, all the mechanical energies and sound that can be captured and converted into electricity using piezoelectric materials. The heat can be captured and converted into electricity using thermal and pyroelectric materials. This book describes basic and advanced concepts of energy harvesting in terms of physics and engineering and then proposes the design techniques to obtain power supplies for low-power systems. The first six chapters describe a special technology of energy harvesting including the different principles of transducer and related materials, power management, storage and design of system. In addition, design techniques with conditioning circuital solutions to efficiently manage a low-power system will be analysed. The final chapter describes various types of energy-harvesting applications and related market with a focus on future architectures.

Pescara, Italy
August 2016

Maurizio Di Paolo Emilio

Acknowledgements

I would like to express my gratitude to all those who gave me the possibility to complete this book. In particular, I want to thank Svetlana Josivofska and Emanuele Bonanni for their foreword and Charles B. Glaser, editorial director, for the publication of present book. To my family, thank you for the patience and for encouraging and inspiring me to follow my dreams. I am especially grateful to my wife, Julia, and my children, Elisa and Federico.

Contents

1 **Introduction** 1
1.1 Fundamentals 1
1.2 Sensors and Transducers 3
1.2.1 Temperature Sensors 3
1.2.2 Magnetic Field Sensors 4
1.2.3 Potentiometers 5
1.2.4 Light Detection 5
1.3 Communications Cabling 7
1.3.1 Noise 7
1.4 Parameters 8
1.4.1 Noise 8
1.4.2 Settling Time 8
1.4.3 DC Input Characteristics 9
References 9

2 **The Fundamentals of Energy Harvesting** 11
2.1 What's Energy? 11
2.2 Why Energy Harvesting? 12
2.3 Free Energy 13
2.4 Power Management Unit 15
2.5 Storage Systems 17
References 19

3 **Input Energy** 21
3.1 Mechanical Energy 21
3.2 Thermal Energy 24
3.3 Electromagnetic Energy 26
3.4 Space Radiation 27
3.5 Solar Radiation 28
3.5.1 Photovoltaic Cell 30
References 33

4 Electromagnetic Transducers ... 37
4.1 Introduction ... 37
4.2 Electromagnetic Waves and Antenna ... 37
4.3 System Design ... 41
References ... 43

5 Piezoelectric Transducers ... 47
5.1 Introduction ... 47
5.2 Materials ... 47
5.3 Model ... 48
5.4 System Design ... 51
References ... 52

6 Thermoelectric Transducers ... 55
6.1 Introduction ... 55
6.2 Seebeck and Peltier Effect ... 55
6.3 Potential ... 57
6.4 Charges in a Semiconductor with a Temperature Gradient ... 57
6.5 Thermoelectric Effect ... 58
6.6 Thomson Effect ... 58
6.7 Thermoelectric Generator ... 59
6.8 Materials ... 60
6.9 Figure of Merit ... 61
References ... 62

7 Electrostatic Transducers ... 65
7.1 Introduction ... 65
7.2 Physical Phenomena ... 66
7.3 Switching System ... 67
7.4 Continuous Systems ... 69
7.5 Design ... 70
References ... 72

8 Powering Microsystem ... 75
8.1 Power Conditioning ... 75
8.2 Rectifier Circuit ... 78
8.2.1 Bridge Rectifier Circuit ... 79
8.2.2 Zener Diode as Voltage Regulator ... 81
8.2.3 Considerations ... 83
8.3 Piezoelectric Biasing ... 84
8.4 Voltage Control ... 86
8.5 MPPT ... 87
8.6 Architecture ... 88
8.7 DC-DC Systems ... 89
8.7.1 Linear Regulators ... 89
8.7.2 Switching Regulators ... 90
8.7.3 Buck Converter ... 91

8.7.4 Boost Converter 91
8.7.5 Buck-Boost Converter 92
8.7.6 Armstrong Oscillator 94
8.8 Load Matching 95
8.9 AC-DC Systems 97
8.10 Electrical Storage Buffer 98
8.10.1 Supercapacitors 100
References 102

9 Low-Power Circuits 105
9.1 Introduction 105
9.2 Review of Microelectronics 105
9.2.1 Basic of Semiconductor's Physics 106
9.2.2 PN Junction 108
9.2.3 Diode 110
9.2.4 Bipolar Transistor: Emitter Follower 111
9.2.5 MOS Transistor 116
9.2.6 Differential Amplifiers 120
9.2.7 Feedback 122
9.2.8 Effects of Feedback 123
9.2.9 Digital CMOS Circuits 124
9.2.10 CMOS Inverter 125
9.2.11 Current Mirror 125
9.2.12 Ideal Current Mirror 127
9.2.13 Current Mirror BJT/MOS 128
9.3 Low-Power MOSFET 129
9.3.1 General Characteristics of a MOSFET 129
9.3.2 Mosfet Power Control 131
9.3.3 Stage of Amplification 131
9.3.4 Common Source 132
9.4 Analog Circuits 132
9.5 Operational Amplifier 134
9.6 Power Supply and Rejection 135
9.7 Low Noise Pre-amplifiers 137
References 139

10 Low-Power Solutions for Biomedical/Mobile Devices 143
10.1 Introduction 143
10.2 Design of Wearable Devices 144
10.3 RF Solutions for Mobile 146
10.3.1 Ferrite Rod Antenna 146
10.3.2 Circular Spiral Inductor Antenna 148
10.3.3 Folded Dipole 149
10.3.4 Microstrip Antenna 149
10.4 Power Management 150

10.5 Ultra-Low Power 2.4 GHz RF Energy Harvesting and Storage System ... 151
References ... 153

11 Applications of Energy Harvesting ... 155
11.1 Introduction ... 155
11.2 Building Automation ... 155
11.3 Environmental Monitoring ... 157
11.4 Structural Health Monitoring ... 157
11.5 Automotive ... 158
11.6 Projects ... 159
11.7 Solar Infrastructure ... 160
11.8 Wind Energy ... 162
11.9 Conclusions ... 162
References ... 163

Index ... 167

Chapter 1
Introduction

1.1 Fundamentals

Data acquisition systems (DAQ) are the main instruments used in laboratory research from scientists and engineers; in particular, for test and measurement, automation, and so on. Typically, DAQ systems are general-purpose data acquisition instruments that are well suited for measuring voltage or current signals. However, many sensors and transducers output signals must be conditioned before that a board can acquire and transform in digital the signal. The basic elements of DAQ are shown in Fig. 1.1 and are:

- Sensors and Transducers
- Field Wiring
- Signal Conditioning
- Data Acquisition Hardware
- Data Acquisition Software
- PC (with operating system).

Transducers can be used to detect a wide range of different physical phenomena such as movement, electrical signals, radiant energy, and thermal, magnetic, or mechanical energy. They are used to convert one kind of energy into another kind. The type of input or output of the transducer used depends on the type of signal detected or process controlled; in other ways, we can define a transducer as a device that converts one physical phenomena into another one. Devices with input function are called Sensors because they detected a physical event that changes according to some events as, for example, heat or force. Instead, device with output function are called actuators and are used in control system to monitor and compare the value of external devices. Sensors and transducers belong to category of transducers.

M. Di Paolo Emilio, *Microelectronic Circuit Design for Energy Harvesting Systems*, DOI 10.1007/978-3-319-47587-5_1

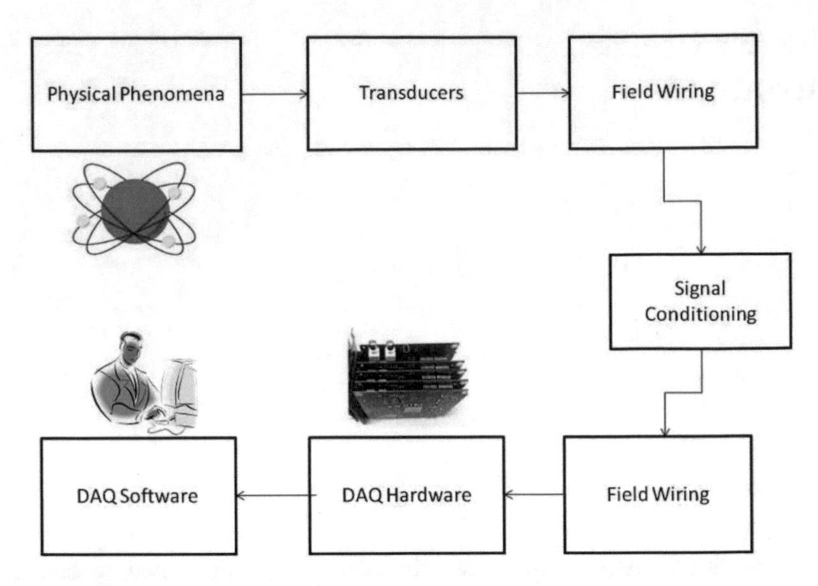

Fig. 1.1 Functional diagram of a PC-based data acquisition system

Table 1.1 Common transducers

Quantity being measured	Input device (Sensor)	Output device (Actuator)
Light level	Photodiode photo-transistor solar cell	Lamps–LED–Fiber optics
Temperature	Thermistor–Thermocouple	Heater–Fan
Force/Pressure	Pressure switch	Electromagnetic vibration
Position	Potentiometer–Encoder	Motor
Speed	Tacho-generator	AC/DC motors
Sound	Carbon microphone	Buzzer–Loudspeaker

There are many different types of transducers; each transducer has input and output characteristics and the choice depends on the goal of your system; for example, from type of signal must be detected and the control system used to manage it (see Table 1.1).

Sensors produce in output a proportional voltage or current signal in according to the variation of physical phenomena that are measuring. There are two types of sensors: active and passive. Active sensors require external power supply to work; instead, a passive sensors generate a signal in output without external power supply. Signal conditioning consists in manage an analog signal in order that it meets the requirements of the next electronics system for additional processing. Generally, in various applications of control system there is a sensing stage (for example, a sensor), conditioning stage, and a processing stage. The conditioning stage can be built, for example, using operational amplifier to amplify the signal and, moreover, can include the filtering, converting, range matching, isolation, and any other processes required to make sensor output suitable for processing stage. The processing stage manages the signal conditioned in other stages such as analog-to-digital converter, microcontroller, and so on [1].

1.2 Sensors and Transducers

Transducers and sensors are used to convert a physical phenomena into an electrical signal (voltage or current) that will then be converted into a digital signal used [2] for the next stage such as a computer, digital system, or memory board.

1.2.1 Temperature Sensors

Several techniques for detection of temperature are currently used. The most common of these are RTDs, thermocouples, thermistors, and sensor ICs. The choice of one for your application can depend on some factors such as required temperature range, linearity, accuracy, cost, and features. Resistance temperature detectors or RTD are more commonly known; they are built using several different materials for the sensing element, for example, the Platinum. Platinum is used for different reasons: high temperature rating, very stable, and very repeatable. Other materials used for RTD sensors are nickel and copper.

Thermocouple is composed of two different metals that have a common contact point where it is produced a voltage (some mV) proportional to the variation of the temperature. Thermistors are generally composed of semiconductor materials. There are thermistors with positive and negative temperature coefficient. The thermistors with negative temperature coefficient are used to monitor low temperature of the order of 10 K [2–4]. The temperature coefficient is defined from the following Eq. (1.1)

$$\alpha(t) = \frac{1}{R(T)}\frac{dR}{dT} \tag{1.1}$$

In general, a linear curve is used working only over a small temperature range. To accurate temperature measurements, it is necessary to use the Steinhart–Hart equation (see (1.2)):

$$\frac{1}{T} = a + b * ln(R) + c * ln^3(R) \tag{1.2}$$

where a, b, and c are parameters. The solution of Eq. (1.2) can be written as (1.3):

$$R = e^{(x-\frac{y}{2})^{\frac{1}{3}} - (x+\frac{y}{2})^{\frac{1}{3}}} \tag{1.3}$$

where

$$x = \sqrt{\left(\frac{b}{3c}\right)^3 + \frac{y^2}{4}} \quad (1.4)$$

and

$$y = \frac{a - \frac{1}{T}}{c} \quad (1.5)$$

Typical values of the resistance of 3000 Ω at room temperature (25 C) are the following:

- $a = 1.40 * 10^{-3}$
- $b = 2.37 * 10^{-4}$
- $c = 9.90 * 10^{-8}$

1.2.2 Magnetic Field Sensors

Magnetic sensors convert magnetic energy into electrical signals for processing by electronic system. Magnetic sensors are designed to respond to a wide range of magnetic field; they are mainly used in different applications, in particular, in automotive systems for the sensing of position, distance, and speed. For example, the position of the car seats and seat belts for air-bag control or wheel speed detection for the anti-lock braking system, (ABS). Magnetic sensors work according to the Hall Effect (see Fig. 1.2): the production of potential difference (Hall Voltage) across a conductor where a perpendicular magnetic field is applied [5–9].

The output voltage, called the Hall voltage, (V_H) of the basic Hall Element is directly proportional to the magnetic field (B) passing through the semiconductor material:

$$V_H = R_H * \left(\frac{I}{t} * B\right) \quad (1.6)$$

where R_H is the Hall Effect coefficient, I is the current flow through the sensor in Ampere, and t is the thickness of the sensor in mm. Most commercial Hall Effect devices are manufactured with built-in DC amplifiers, voltage regulators to improve the sensors sensitivity, and the range of output voltage that it is quite small, only few microvolts [2, 10–15].

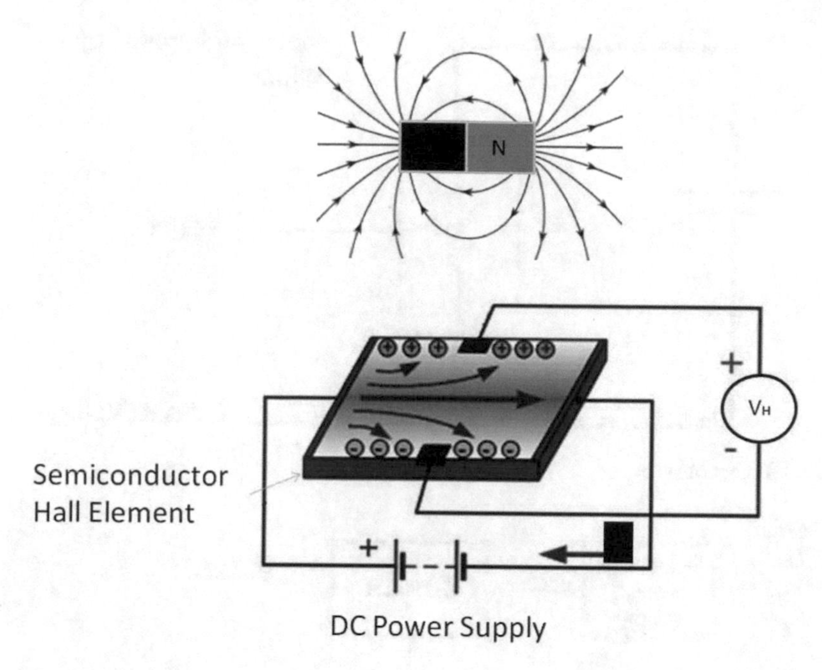

Fig. 1.2 Hall Effect sensor

1.2.3 Potentiometers

A potentiometer is an electromechanical device that contains a movable wiper arm with the goal of maintaining electrical contact with a resistive surface; the wiper is coupled mechanically to a movable linkage. It gives a voltage signal by divider circuit when voltage is applied across the entire resistance within the potentiometer, see Fig. 1.3. A variable potential difference can then be produced at a central wiper arm relative to one of the resistor as the wiper is moved. The wiper is usually made of a material such as beryllium [1].

1.2.4 Light Detection

Light sensors detect light emitted or given off from an object: such as LED, reflected from surfaces, transmitted from electronics device, and so on. LED or light emitting diode, is a solid-state semiconductor that emits light when current through it in the forward direction. A photoelectric (see Fig. 1.4) sensor is an electrical device that responds to the change in the intensity of the light falling upon it [16–22].

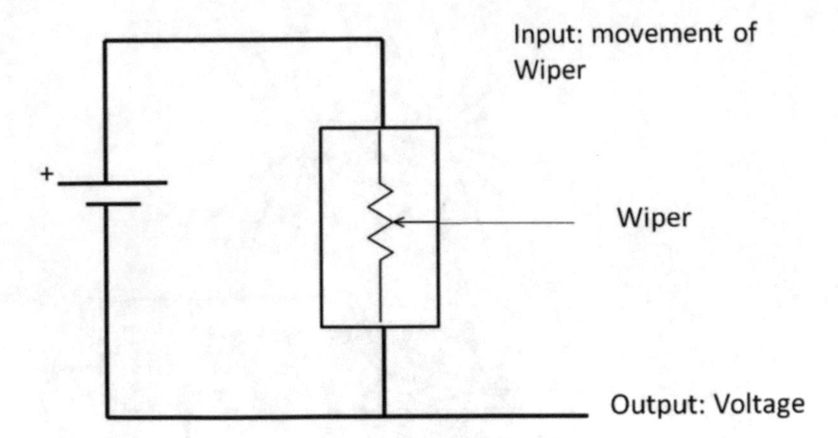

Fig. 1.3 Potentiometers

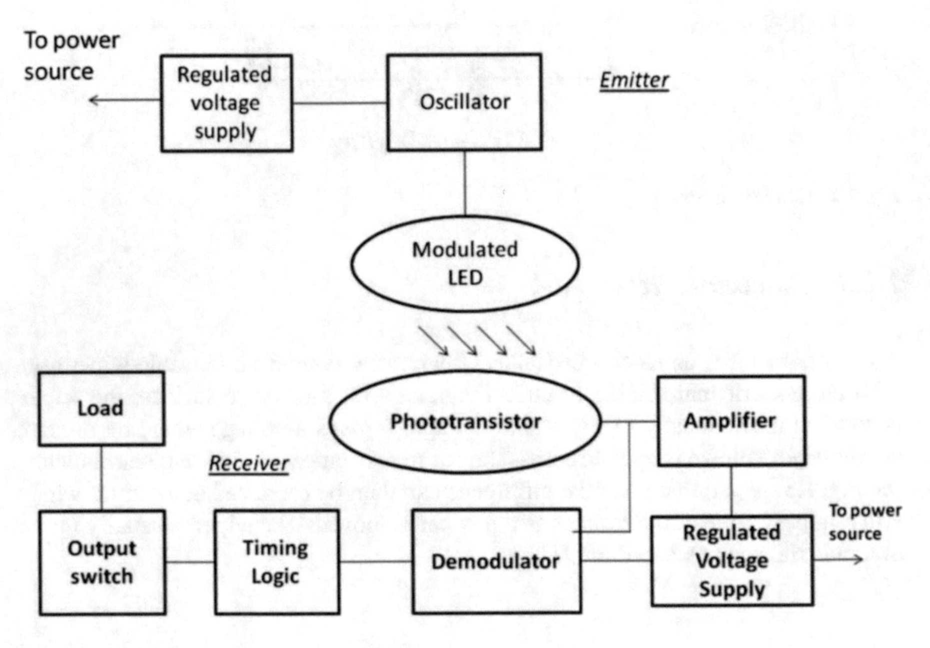

Fig. 1.4 Light detection: data acquisition system

There are many sensing situations where space is too restricted or the environment too hostile even for remote sensors. Fiber optics is an alternative technology in sensor "packaging" for such applications such as photoelectric sensing technology. Moreover, fiber optics are flexible, transparent fiber made of glass (silica). It works as a waveguide to transmit light [1].

1.3 Communications Cabling

Field wiring is the physical connection from the transducers/sensors to the hardware. When the signal conditioning and/or data acquisition hardware is remotely located from the PC/devices, then it is necessary to use field wiring that provides the physical link. In this case, it is very important to estimate the effects of the external noise, especially in industrial environments. In the next paragraph it provides an estimation of this noise [1].

1.3.1 Noise

One characteristics of all electronics circuits is represented of noise: it is a random fluctuation in an electrical signal generated by electronic devices. In communication systems, the noise is an undesired random disturbance of a useful information signal.

1.3.1.1 Thermal Noise

Johnson–Nyquist noise or thermal noise is generated by the random thermal motion of electrons. Thermal noise is approximately a white noise: the amplitude of the signal can be described by a Gaussian probability density function.

The root mean square (RMS) voltage due to thermal noise v_n, generated in a resistance R (ohms) over bandwidth Δf (hertz), is given by

$$v_n = \sqrt{4k_B T R \Delta f} \tag{1.7}$$

where k_B is Boltzmann's constant (joules per kelvin) and T is the resistor's absolute temperature (kelvin).

1.3.1.2 Shot Noise and Flicker Noise

Shot noise in electronic devices consists of unavoidable random statistical fluctuations of the electric current in an electrical conductor. Moreover, Flicker noise, also known as $1/f$ noise, occurs in almost all electronic devices, and results from a variety of effects, though always related to a direct current [1, 23–25].

1.4 Parameters

To properly design a data acquisition system, we must know some important parameters. The goal of this section is to describe major system parameters for a better design in various field of the electronics, in particular, data acquisition systems, microelectronics, and in the power management system for energy harvesting [23].

1.4.1 Noise

Each measurement generates noise as a combination of more signals. It is the interference between two terminals. One factor, common-mode noise, indicates the interferences that appear on both measurements inputs. The majority of common-mode interference is attributable to 50 Hz (or 60 Hz) power frequency [26, 27].

1.4.2 Settling Time

The settling time of an electronic device is the time elapsed from the application of an ideal step input to the time at which the value output has entered and remained within a specified error range. Parameters that can describe settling time are the following: propagation delay and time required for obtain output value (Fig. 1.5) [23, 28, 29].

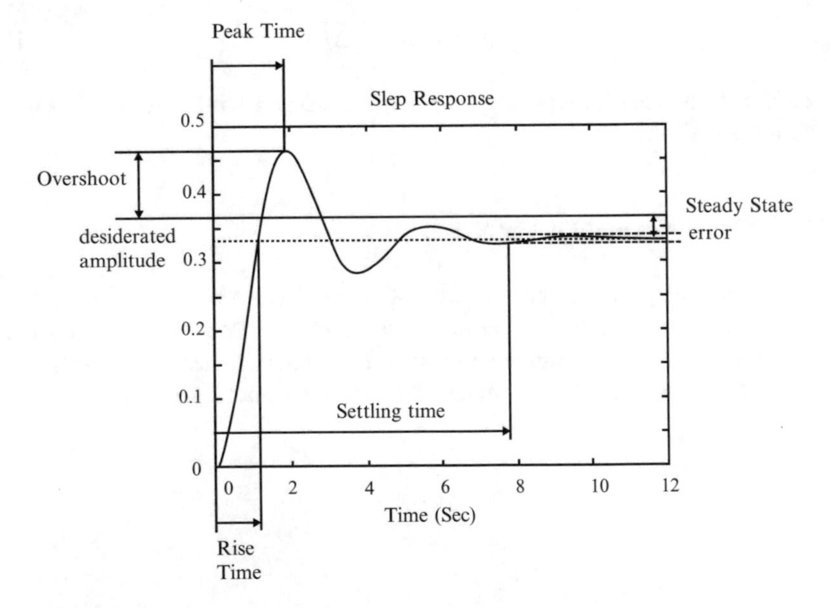

Fig. 1.5 Settling time

1.4.3 DC Input Characteristics

It indicates the value of offset voltages, offset currents, and bias current of electronic devices [30–35].

References

1. Park, J., & Mackqy, S. (2003). *Practical data acquisition for instrumentation and system control* Elsevier, Oxford.
2. Lacanette, K. (2003). *National temperature sensors handbook.* Annali di Matematica Pura ed Applicata. National Semiconductor.
3. National Instruments. (1996). *Data acquisition fundamentals,* Application note 007.
4. National Instruments. (1996). *Signal conditioning fundamentals for PC-based data acquisition,* Application Note 048
5. Roundy, S. (2003). *Energy scavenging for wireless sensor nodes with a focus on vibration to electricity conversion.* Ph.D Thesis, University of California.
6. Tsutsumino, T., Suzuki, Y., Kasagi, N., Kashiwagi, K., & Morizawa, Y. (2006). *Micro seismic electret generator for energy harvesting.* Technical Digest PowerMEMS (pp. 133–136). Berkeley, USA.
7. Sterken, T., Altena, G., Fiorini, P., & Puers, R. (2007). *Characterisation of an electrostatic vibration harvester, EDA Publishing Association.*
8. Sterken, T., Baert, K., Puers, R., & Borghs, S. (2002). Power extraction from ambient vibration. In *Proceedings of the SeSens (Workshop on Semiconductor Sensors, Veldhoven, Netherlands)* (pp. 680–683).
9. Szarka, G., Stark, B., & Burrow, S. (2012). Review of power management for energy harvesting systems. IEEE *Transactions on Power Electronics, 27*(2), 803–815. ISSN: 0885-8993.
10. Cammarano, A., Burrow, S. G., Barton, D. A. W., Carrella, A., & Clare, L. R. (2010). Tuning a resonant energy harvester using a generalized electrical load. *Journal of Smart Materials and Structures, 19,* 055003.
11. Guyomar, D., Badel, A., Lefeuvre, E., & Richard, C. (2005). Toward energy harvesting using active materials and conversion improvement by nonlinear processing. *IEEE Transactions on Ultrasonics, Ferroelectrics, and Frequency Control, 52,* 584–595.
12. Mitcheson, P. D., Stoianov, I., & Yeatman, E. M. (2012). Power-extraction circuits for piezoelectric energy harvesters in miniature and low-power applications. *IEEE Transactions on Power Electronics, 27,* 4514–4529.
13. Szarka, G. D., Burrow, S. G., & Stark, B. H. (2012). Ultra-low power, fully-autonomous boost rectifier for electro-magnetic energy harvesters. *IEEE Transactions on Power Electronics, 28*(7), 3353–3362. doi:10.1109/TPEL.2012.2219594.
14. Maurath, D., Becker, P. F., Spreeman, D., & Manoli, Y. (2012). Efficient energy harvesting with electromagnetic energy transducers using active low-voltage. *IEEE Journal of Solid-State Circuits, 47*(6)
15. Beeby, S. P., Tudor, M. J., & White, N. M. (2006). Energy harvesting vibration sources for microsystems applications. *Measurement Science and Technology, 17,* R175–R195.
16. Khaligh, A., Zeng, P., & Zheng, C. (2010). Kinetic energy harvesting using piezoelectric and electromagnetic technologies—state of the art. *IEEE Transactions on Industrial Electronics, 57*(3), 850–860.
17. Paulo, J., & Gaspar, P. D. (2010). Review and future trend of energy harvesting methods for portable medical devices. In *Proceedings of the World Congress on Engineering* (Vol. 2)

18. Zhu, D., Tudor, M. J., & Beeby, S. P. (2010). Strategies for increasing the operating frequency range of vibration energy harvesters: A review. *Measurement Science and Technology, 21*, 022001-1–022001-29.
19. Cepnik, C., Lausecker, R., & Wallrabe, U. (2013). Review on electrodynamic energy harvesters—a classification approach. *Micromachines, 4*(2), 168–196. http://www.mdpi.com/2072-666X/4/2/168. Accessed 20 Jan 2015.
20. Ulaby, F. T., Michielssen, E., & Ravaioli, U. (2010). *Fundamentals of applied electromagnetics* (6th ed.). Prentice Hall.
21. Roundy, S., Wright, P. K., & Rabaey, J. M. (2003). A study of low level vibrations as a power source for wireless sensor nodes. *Computer Communications, 26*(11), 1131–1144.
22. Sazonov, E., Li, H., Curry, D., & Pillay, P. (2009). Self-powered sensors for monitoring of highway bridges. *IEEE Sensors Journal, 9*, 1422–1429.
23. Taylor, J. (1986). *Computer-based data acquisition system*. Instrument Society of America, USA.
24. Di Paolo Emilio, M. (2013). *Data acquisition system, from fundamentals to applied design*. New York: Springer.
25. Roundy, S., Wright, P., & Pister. K. (2002). Micro-electrostatic vibration-to- electricity converters. In *Proceedings of ASME international mechanical engineering congress and exposition (IMECE)* (Vol. 220, pp. 17–22).
26. Stordeur, M., & Stark, I. (1997). Low power thermoelectric generator: Self-sufficient energy supply for micro systems. In *Proceedings of the 16th international conference on thermoelectrics* (pp. 575–577).
27. Shenck, N., & Paradiso, J. (2001). Energy scavenging with shoe-mounted piezoelectrics. *Micro IEEE, 21*(3), 30–42.
28. Toh, T. T., Mitcheson, P. D., Holmes, A. S., & Yeatman, E. M. (2008). A continuously rotating energy harvester with maximum power point tracking. *Journal of Micromechanics and Microengineering, 18*, 104008-1-7.
29. Howey, D. A., Bansal, A., & Holmes, A. S. (2011). Design and performance of a centimetre-scale shrouded wind turbine for energy harvesting. *Smart Materials and Structures, 20*, 085021.
30. Razavi, B. (2002). *Design of analog CMOS integrated circuits*. McGraw-Hill
31. Razavi, B. (2008). *Fundamentals of microelectronics*. New York: Wiley.
32. Sedra, A. S., & Smith, K. C. (2013). *Microelectronic circuits*. Oxford: Oxford University.
33. Razavi, B. (2002). *Design of integrated circuits for optical communications*. McGraw-Hill.
34. Hurst, P. J. (2001). *Analysis and design of analog integrated circuits*. New York: Wiley.
35. Spies, P. (2015). *Handbook of energy harvesting power supplies and applications*. CRC Press Book, France.

Chapter 2
The Fundamentals of Energy Harvesting

2.1 What's Energy?

Energy is the capacity to do work; in the physics field, work is something resulting from the action of a force such as that of gravity. In Nature there are different types of energy: the most classic case is solar energy and all that energies come from the universe in the form of cosmic rays, X-rays, gravitational waves, dark matter, etc. A system that produces energy can be represented from a kite that "floats" in the clouds by means of wind, or a wave of light is passing through a space. According to the energy conservation law, for example, one of the first laws of thermodynamics, the total energy of a system is conserved, although it can be transformed into another form. Two billiard balls can collide, for example, and energy transformations are involve with sound and heat at the contact point: this phenomena is derived from energy conservation law after the collision. In few words, all forms of energy can be converted into another. This had already begun when the man (or woman) lit the first fire by burning wood with the transformation of the chemical energy of the molecules in the form of heat. The energy transfer is based on energy conservation. Other examples, a battery that generates electrons from chemical reactions, a toaster, the automobile, and many others. The sound is a form of kinetic energy: it is caused from vibration of the air molecules described as mathematical models. This vibration energy is transformed into electrical pulses that can be interpreted from the human as sound wave. In some systems such as that for the production of nuclear energy, the atoms are involved in multiple processes: the atoms of the nuclear fuel are divided by releasing the creation of thermal energy which is capture as water vapor to drive a kinetic energy generator. Subsequently, a motor turns it into a current flow to provide power supply. Renewable energy (replenished naturally) is generated from natural sources such as sunlight, wind, rain, tides, and geothermal heat, which are renewable (that are replenished naturally). Alternative energy is a term used for an energy source that is an alternative to the use of fossil fuels with a low environmental impact [1–6]. In the International System of units (SI), the unit

M. Di Paolo Emilio, *Microelectronic Circuit Design for Energy Harvesting Systems*, DOI 10.1007/978-3-319-47587-5_2

of energy is the joule, named thanks to James Prescott Joule. It is a derived unit and matched to the energy expenditure (or work) by applying a force of one Newton for a distance of one meter. However, energy is also expressed in many other units that are not part of the SI, as ergs, calories, British Thermal Units, kilowatt hours, and kilocalories, they require a conversion factor when expressed in SI units. In classical mechanics, from a mathematical point of view, energy is a conserved quantity. The work, a form of energy, is a force over a given distance described by the following equation:

$$w = \int_C Fds \tag{2.1}$$

Eq. (2.1) tells us that the work is equal to the line integral of the force along a path C. In the energy field some terms are used: Hamiltonian and Lagrangian. The total energy of a system can be expressed by Hamiltonian by using motion equation of William Rowan Hamilton. Another energy-related concept is called Lagrange, from Joseph-Louis Lagrange. This formalism is mathematically more convenient than the Hamiltonian for non-conservative systems (such as friction systems). The Lagrange is defined as the kinetic energy minus the potential energy.

2.2 Why Energy Harvesting?

An energy harvesting system captures the environmental energy and converts it into electricity. There are many techniques about this field: for example, to capture the energy lost or dissipated as heat, light, sound, vibration, or movement, then using special electronic circuits to manage the collection energy and then transform it into electrical signal.

This is important, because our existing electrical infrastructure is extremely wasteful in its use of energy. For example, some today's technologies used in the production of electricity are not energy efficient. The old incandescent bulbs, now they are no longer sold, transform into heat a good percentage of electricity.

An energy harvesting system can be described as in Fig. 2.1. The blocks are described in the following points:

- Energy transducer used to convert ambient energy into electrical energy of input. Environmental sources of energy available for the conversion may be the following: heat (thermoelectric modules), light (solar cells), RF radiation (antennas), and vibration (piezoelectric).
- Rectifier and super capacitor: a rectifier and an optional storage system for energy management.
- Voltage regulator: a controller system for adapting the voltage level to the requirements of the powered device.

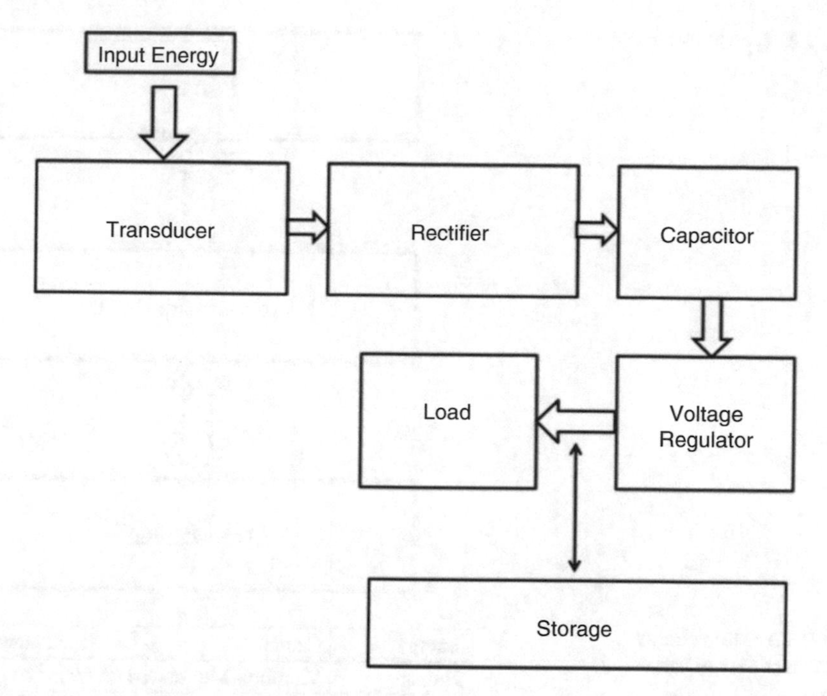

Fig. 2.1 General system of energy harvesting

- Optional energy storage element: depending on the requirements of application, it is possible to use a battery as an energy storage element. In some systems it can be activated at intervals of time, in others it is powered (or recharged) permanently.
- Load: the impedance of the system to power. It may have different ways of energy consumption making the whole system work in low-power mode.

The electrical energy obtained from an energy harvesting system is very small (about 1 W/cm^3 to 100 mW/cm^3) and therefore has a working point in low-power mode. A typical electronic load is constituted by a sensor, a microcontroller, and a wireless transceiver (Fig. 2.2). Energy consumption is about of uA for the first two components and a few mA for the transceivers. These considerations will be used in the design of the energy harvesting system.

2.3 Free Energy

The main energy sources "freely used" are solar, mechanical, and thermal. The self-powered devices are normally of small dimensions that belong to the category of wearable devices or otherwise forming part of the Internet of Things (IoT) system. A possible comparison can be made in terms of power density per unit volume. In Table 2.1 are summarized the main sources of energy [7–13].

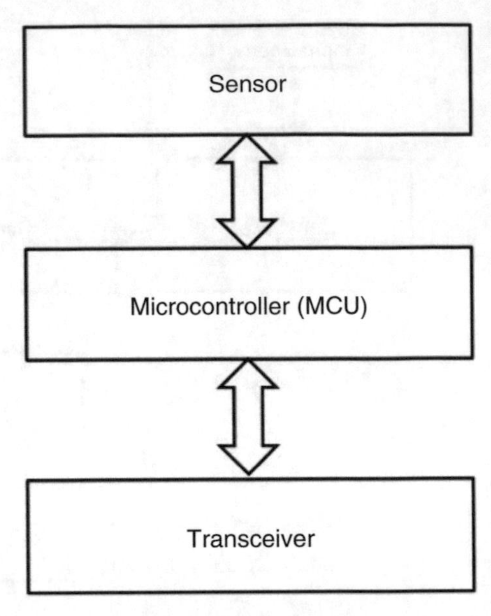

Fig. 2.2 Typical electronic load

Table 2.1 Main energy sources of energy harvesting

Energy	Category	Harvested power
Human	Vibration/Motion	$4\,\mu W/cm^2$
Industry	Vibration/Motion	$100\,\mu W/cm^2$
Human	Temperature	$25\,\mu W/cm^2$
Industry	Temperature	$1–10\,\mu W/cm^2$
Indoor	Light	$10\,\mu W/cm^2$
Outdoor	Light	$10\,mW/cm^2$
GSM/3G/4G	RF	$0.1\,\mu W/cm^2$
Wi-Fi	RF	$1\,\mu W/cm^2$

The light is a source of ambient energy available for low and high power electronic devices. A photovoltaic system generates electricity by converting the solar for a large number of applications in the various range of power. The sunlight varies on the surface of the earth, depending on weather conditions and the position expressed in terms of longitude and latitude. For each position there is an optimal tilt angle and orientation of the solar cells in order to obtain the maximum radiation for powering high power systems; these conditions are not suitable in the case of small solar cells for the wearable electronics where there are not landmarks and the design is done in according to the general case [14–20]. The sun radiates towards the earth's surface with a power density of at least $1350\,J/m^2$, with a total power on earth about $170 * 10^9$ MW.

As in almost all the transformations, kinetic energy is the base of the harvesting in terms of movement of particles such as photons (sun) or generic waves. The movement or deformation is converted into electrical energy in three main modes: inductive, electrostatic, and piezoelectric [21–26].

Table 2.2 Vibration sources

Source	Peak acceleration (m/S^2)	Frequency (Hz)
HVAC vents	0.2–1.5	4 50/60
Microwave oven	2.3	121
Dryer	3.5	121
Notebook with CD/DVD	0.6	75
Washing machine	0.5	109

The vibrations is the energy source for mechanical transducers and are characterized by two parameters: acceleration and frequency. Table 2.2 visualizes a list of peak accelerations and frequencies for different vibration sources in the industrial field. From these data can be noted that the vibrations of industrial machines have accelerations between 60 and 125 Hz. There is another possibility to use human body as a source of vibrations. The vibrations associated with the human body have accelerations with frequencies below about 10^8 Hz.

The Human Walking, for example, is one of the activities that have more energy associated for the production of electrical signals. Two power modes can be distinguished: active and passive. The active power of electronic devices occurs when the user needs to do a specific work to power the device. The passive mode, instead, is when the humans must not do any works than their daily activities: finger movement, walking, heat of the body, etc.

2.4 Power Management Unit

The piezoelectric modules and the transducers operate with an output voltage of the order of mV, which changes in according to the environmental conditions and materials used. The electronic circuits, such as microcontrollers or wireless transceivers which are very often used in power supplies energy harvesting, generally work with a supply voltage of between 1.8 and 5 V. They need constant power to maximize their performance. The ripple oscillations declass the performance in terms of parameters such as noise figure and accuracy. The property to suppress that noise in a power supply line is expressed by the PSRR. PSRR is introduced to indicate the amount of noise introduced from a power supply, it stands for "Power Supply Rejection Ratio" expressed in *dB* as the ratio of the variation of the supply voltage in an operational amplifier and the equivalent output voltage (differential). PSRR is the main design parameter in modern SoC [27–32]. In order to limit the oscillations, different circuits of power management such as boost converter are used. The problem of the threshold voltage is reflected in the possibility of not power a circuit if it doesn't exceed 0.3 V. Various techniques (star-up circuits) are used during the startup when the battery is not present in the system. After a level transition, as soon as the converter provides stable voltage to the circuit, the start-up circuits are disabled. These circuits are used primarily with thermogenerator.

Another important aspect is the system impedance. With a certain power, the source must provide maximum power for a given load. To adapt the impedance, Tracker MPPT circuits such as those employed in the photovoltaic panels are used. In the switching regulators, frequency variation is reflected in an input impedance change. In this way, the tracker controls the input resistance to achieve maximum transducer power. For a piezoelectric, to extract more energy implies a perfect layout of power management. A transceiver requires a certain time and then a current to run a data transmission in a specific time. These considerations can be expressed as the following:

$$T_a = \frac{1}{d_r * \frac{1}{\frac{D}{n} * m}} \tag{2.2}$$

where d_r is the data flow of data rate (bytes/s), D are the bytes of data to be transmitted, n are the bytes of a data package, and m is the length of the package. In the procedure for sending and receiving data, all necessary blocks are activated unlike the sleep mode. The average current required by the transceiver is the following:

$$I = \frac{I_{\text{sleep}} T_{\text{sleep}} + I_T T_T}{T_s} \tag{2.3}$$

where I_{sleep} is the current absorbed by the transceiver in the sleep mode with the corresponding sleep time, I_T is the current consumed during a transmission time T_T, and T_s is the transmission period which is the sum of the sleep and transmission time. The important parameters in the selection of a sensor for energy harvesting applications are the current consumption in both active and passive mode, the average power and the sleep time. Sensors can provide an analog or digital output. The I2C/SPI bus provides a direct interface for transmitting data to digital circuits such as microcontroller. The sensitivity of the sensor is the amount of variation of the output signal in according to the variation of the measured parameter. A conditioning circuit to manage the output voltage before being sent to a control device is required. The response bandwidth of a sensor is expressed in Hertz and is the maximum rate at which the sensor can work correctly. The microcontrollers have different working operation mode in according to the current consumption. In the active mode the consumption is obviously higher than passive mode with some circuit parts turned off. In the active mode of low-power microcontrollers, the current consumption is higher than passive mode and all the clocks are active, while in the low-power consumption mode the CPU and some of the internal clocks are disabled. In Fig. 2.3 is visualized a generic block diagram of a low-power microcontroller, while Fig. 2.4 shown a typical current profile for a wireless transceiver. In an general energy harvesting system, the transceiver is in standby mode in most of the time to keep the average power consumption to a minimum. When the data are to be transmitted, the transmission mode is activated and then the maximum peak current is consumed [33–35].

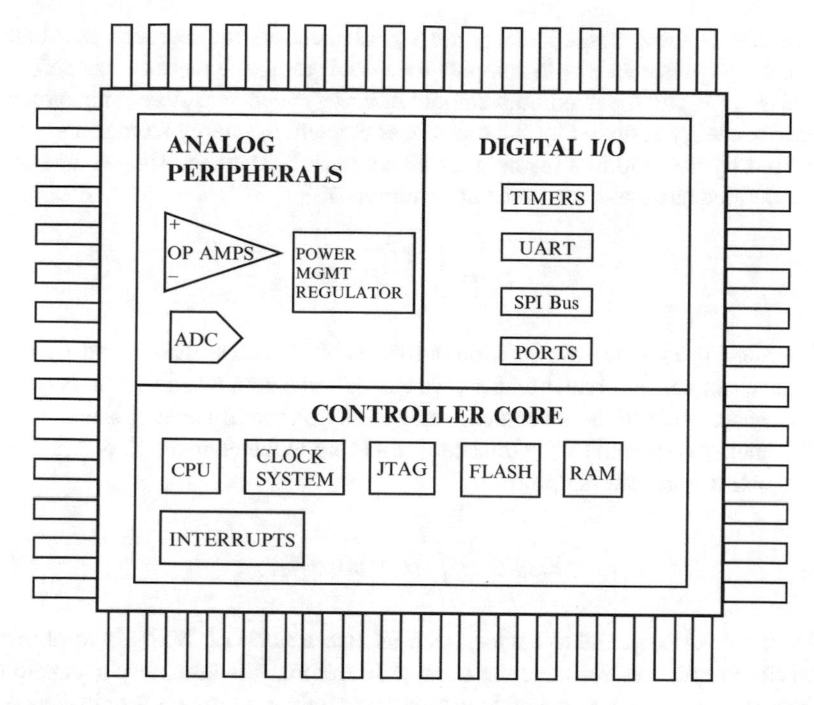

Fig. 2.3 Block diagram of a low-power microcontroller

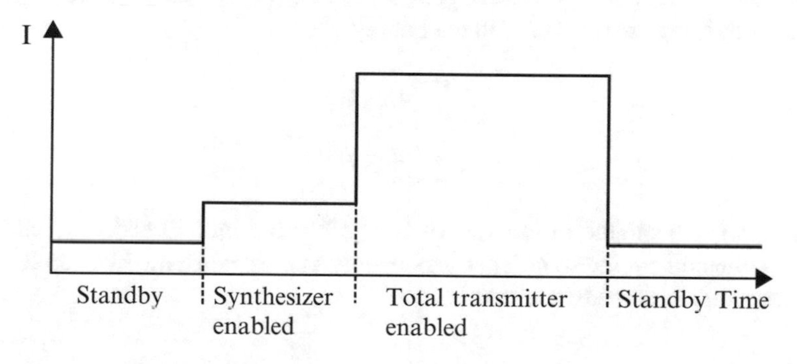

Fig. 2.4 Typical current waveform for a wireless transceiver

2.5 Storage Systems

In an energy harvesting system, the output voltage is not constant in time. It's very important to have the same amount of constant energy over time, a bit as in photovoltaic devices. That means have the same average power in the time interval in according to load system. An energy storage element is not necessary if the power consumption of the electronic device is always less than the power generated by energy harvesting generator that is actuated only when there is power generated.

For the rest of the cases, an energy storage element is required such as a battery. The goal is to present a way to compute the initial storage element charge before the commissioning and the maximum amount of energy needed to store. First, you must define the energy supplied by the transducer element of energy storage and energy consumed by the load in a mathematical way must be defined. The average power can be defined as measuring in the time interval T:

$$p_s = \frac{1}{T}\int_0^T P_s(t)dt \tag{2.4}$$

Considering T_{l-i} the time duration during which P_s is less than p_s and T_{h-l} is the opposite case, i.e., P_s greater than p_s, we can define a deficit (minimum) energy for the first case, while in the second case an excess (maximum) energy is defined.

The energy collected by the transducer and sent to the element of energy storage will be within a certain margin:

$$E_{\min} \leq \frac{1}{T}\int_0^T P_s(t)dt \leq E_{\max} \tag{2.5}$$

The goal for a wearable device, such as also a node of WSN, is to eliminate the need to replace or recharge the battery. Therefore, it is necessary to ensure that the battery is always maintained to the energy required by the electronic device. To ensure this, the total energy must be greater than that consumed by the load, in other words (with B the energy stored in the battery):

$$\sum E > 0 \tag{2.6}$$

$$\sum E \leq B \tag{2.7}$$

The energy available in the system is equal to the initial energy stored in the energy collected by the transducer less battery energy consumed by the load and any losses due to parasitic factors:

$$\sum E = B + E_s - El - \int_0^T P_{\text{leak}}Tdt \tag{2.8}$$

where P_{leak} is the leakage power of the storage element. In the case of a load with different power consumption modes as displayed in Fig. 2.4, the calculation of the required power is done by considering the following equation:

$$\tau = \sum_{i=1}^{N} T_i\tau \tag{2.9}$$

T_i is the time interval where the consumption power is P_i. Generally the following equation can be defined:

$$p_l = \sum P_i T_i \tag{2.10}$$

To acquire data from remote locations, the sensor nodes designed for the Internet of Things (IoT) must be able to function for as long as possible on a single battery charge. In an ideal approach, it would not need a battery because its existence can complicate the management of the system. One major problem is that the power is difficult to "catch," it comes as a very low level but with phase problems to resolve. Accordingly, specialized techniques are required for the inputs, which include a boost converter capable of handling the low-voltage sources, high impedance, and other characteristics of many energy harvesting modules. Furthermore, circuits such as boost converters may introduce high frequency noise that may disturb radio communications.

References

1. Park, J., & Mackqy, S. (2003). *Practical Data Acquisition for instrumentation and system control*. Elsevier, Oxford.
2. Lacanette, K. (2003). *National temperature sensors handbook*. Annali di Matematica Pura ed Applicata. National Semiconductor.
3. National Instruments. (1996). *Data acquisition fundamentals*, Application note 007.
4. National Instruments. (1996). *Signal conditioning fundamentals for PC-based data acquisition*, Application Note 048
5. Taylor, J. (1986). *Computer-based data acquisition system*. Instrument Society of America.
6. Di Paolo Emilio, M. (2013). *Data acquisition system, from fundamentals to applied design*. New York: Springer.
7. Roundy, S., Wright, P., & Pister. K. (2002). Micro-electrostatic vibration-to- electricity converters. In *Proceedings of asme international mechanical engineering congress and exposition (IMECE)* (Vol. 220, pp. 17–22).
8. Stordeur, M., & Stark, I. (1997). Low power thermoelectric generator: self-sufficient energy supply for micro systems. In *Proceedings of the 16th international conference on thermoelectrics* (pp. 575–577).
9. Shenck, N., & Paradiso, J. (2001). Energy scavenging with shoe-mounted piezoelectrics. *Micro IEEE 21*(3), 30–42.
10. Roundy, S. (2003). *Energy scavenging for wireless sensor nodes with a focus on vibration to electricity conversion*. Ph.D Thesis, University of California.
11. Tsutsumino, T., Suzuki, Y., Kasagi, N., Kashiwagi, K., & Morizawa, Y. (2006). *Micro seismic electret generator for energy harvesting*. Technical Digest PowerMEMS (pp. 133–136). Berkeley, USA.
12. Sterken, T., Altena, G., Fiorini, P., & Puers, R. (2007). *Characterisation of an electrostatic vibration harvester*, EDA Publishing Association.
13. Sterken, T., Baert, K., Puers, R., & Borghs, S. (2002). Power extraction from ambient vibration. In *Proceedings of the SeSens (Workshop on Semiconductor Sensors, Veldhoven, Netherlands)* (pp. 680–683).

14. Szarka, G., Stark, B., & Burrow, S. (2012). Review of power management for energy harvesting systems. *IEEE Transactions on Power Electronics, 27*(2), 803–815. ISSN: 0885-8993.
15. Cammarano, A., Burrow, S. G., Barton, D. A. W., Carrella, A., & Clare, L. R. (2010). Tuning a resonant energy harvester using a generalized electrical load. *Smart Materials and Structures, 19*, 055003.
16. Guyomar, D., Badel, A., Lefeuvre, E., & Richard, C. (2005). Toward energy harvesting using active materials and conversion improvement by nonlinear processing. *IEEE Transactions on Ultrasonics, Ferroelectrics, and Frequency Control, 52*, 584–595.
17. Mitcheson, P. D., Stoianov, I., & Yeatman, E. M. (2012). Power-extraction circuits for piezoelectric energy harvesters in miniature and low-power applications. *IEEE Transactions on Power Electronics, 27*, 4514–4529.
18. Szarka, G. D., Burrow, S. G., & Stark, B.H. (2012). Ultra-low power, fully-autonomous boost rectifier for electro-magnetic energy harvesters. *IEEE Transactions on Power Electronics, 28*(7), 3353–3362. doi:10.1109/TPEL.2012.2219594.
19. Maurath, D., Becker, P. F., Spreeman, D., Manoli, Y. (2012). Efficient energy harvesting with electromagnetic energy transducers using active low-voltage. *IEEE Journal of Solid-State Circuits, 47*(6), 1369–1380
20. Beeby, S. P., Tudor, M. J., & White, N. M. (2006). Energy harvesting vibration sources for microsystems applications. *Measurement Science and Technology, 17*, R175–R195.
21. Khaligh, A., Zeng, P., & Zheng, C. (2010). Kinetic energy harvesting using piezoelectric and electromagnetic technologies—state of the art. *IEEE Transactions on Industrial Electronics, 57*(3), 850–860.
22. Paulo, J., & Gaspar, P. D. (2010). Review and future trend of energy harvesting methods for portable medical devices. In *Proceedings of the World Congress on Engineering* (Vol. 2)
23. Zhu, D., Tudor, M. J., Beeby, S. P. (2010). Strategies for increasing the operating frequency range of vibration energy harvesters: A review. *Measurement Science and Technology, 21*, 022001-1–022001-29.
24. Cepnik, C., Lausecker, R., & Wallrabe, U. (2013). Review on electrodynamic energy harvesters—a classification approach. *Micromachines, 4*(2), 168–196. http://www.mdpi.com/2072-666X/4/2/168. Accessed 20 Jan 2015.
25. Ulaby, F. T., Michielssen, E., & Ravaioli, U. (2010). *Fundamentals of Applied Electromagnetics* (6th ed.). Prentice Hall, USA.
26. Roundy, S., Wright, P. K., & Rabaey, J. M. (2003). A study of low level vibrations as a power source for wireless sensor nodes. *Computer Communications, 26*(11), 1131–1144.
27. Sazonov, E., Li, H., Curry, D., & Pillay, P. (2009). Self-powered sensors for monitoring of highway bridges. *IEEE Sensors Journal, 9*, 1422–1429.
28. Toh, T. T., Mitcheson, P. D., Holmes, A. S., & Yeatman, E. M. (2008). A continuously rotating energy harvester with maximum power point tracking. *Journal of Micromechanics and Microengineering, 18*, 104008-1-7.
29. Howey, D. A., Bansal, A., & Holmes, A. S. (2011). Design and performance of a centimetre-scale shrouded wind turbine for energy harvesting. *Smart Materials and Structures, 20*, 085021.
30. Razavi, B. (2002). *Design of analog CMOS integrated circuits*. McGraw-Hill
31. Razavi, B. (2008). *Fundamentals of microelectronics*. New York: Wiley.
32. Sedra, A. S., & Smith, K. C. (2013). *Microelectronic circuits*. Oxford: Oxford University.
33. Razavi, B. (2002). *Design of integrated circuits for optical communications*. McGraw-Hill.
34. Hurst, P. J. (2001). *Analysis and design of analog integrated circuits*. New York: Wiley.
35. Spies, P. (2015). *Handbook of energy harvesting power supplies and applications*. CRC Press Book, France.

Chapter 3
Input Energy

3.1 Mechanical Energy

The possibility of avoiding the replacement of the batteries is very attractive especially for the network WSN sensors, where the maintenance costs are high. Other fields of application is the biomedical field where through piezoelectric sensors can be implement touchable sensors. Recent research also includes the conversion of energy from the occlusal contact during mastication by means of a piezoelectric layer and the heartbeat (Fig. 3.1).

The mechanical energy is present in nature in the form of various examples, for example, in vibrating structures or fluid is flowing along the structures. Wherever there is a mass there is a high potential for energy harvesting applications. The resulting energy must be oscillating in a periodic shape as those of a motor, or causal as in most of the natural phenomena. Simple examples of energy harvesting by using fluids are represented by windmills: exploiting the fluid mechanical energy, electricity is generated by electromechanical generators. The main source of mechanical energy for energy harvesting is vibrational. Any system, whatever it is, is subjected to vibrations that somehow can be harvested to generate electricity. Simple examples are body movements, as well as those of animals or other vibrations resulting from movement of building structures [1–6]. In mechanical vibrations, the mathematical expression for the state variable of $q(t)$ varies in time following a harmonic motion: the latter is the most simple representation described with trigonometric functions. The variable $q(t)$ represents a displacement, force, or pressure angle, and can assume the following expression:

$$q(t) = |A| \cos(\omega t + \phi) \tag{3.1}$$

where $|A|$ is the amplitude of the motion, ω the frequency in radians per second, and ϕ is the phase angle. To characterize a motion, all parameters are important: first of all not only the amplitude but also the frequency that identifies the band

M. Di Paolo Emilio, *Microelectronic Circuit Design for Energy Harvesting Systems*, DOI 10.1007/978-3-319-47587-5_3

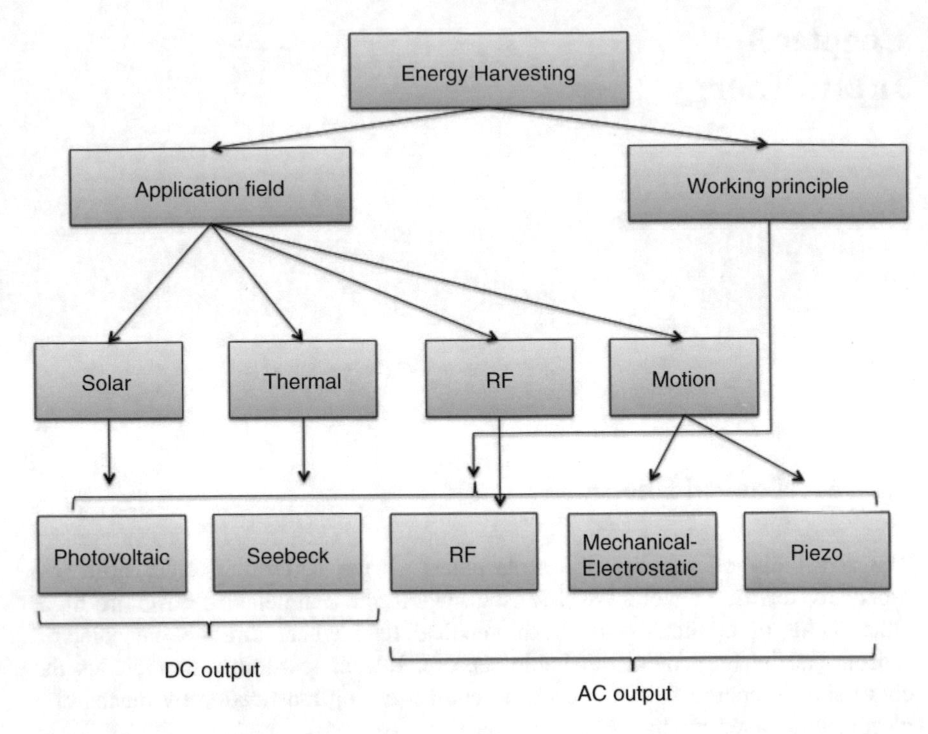

Fig. 3.1 Sources of energy for harvesting

of frequencies with one dominant and the other of a higher order. The vibrations excited by humans are under the 2 Hz; those between 20 Hz and 20 kHz are audible to the human ear and therefore are more likely to be caught. The lowest natural frequency is called the fundamental frequency or dominant. It is useful to know the natural frequencies of a structure to tune an energy harvesting system exactly at a given frequency. The transducers measure the movement and turn it into an electric signal in the time domain, and then with an FFT in the frequency domain. The frequency analysis allows to determine the various frequencies and damping of the system parameters. One or more transducers of movement measure the output vibration. The signals measured in the time domain are then transformed into the frequency domain, and modal analysis calculates the natural frequencies, eigenvectors, and damping parameters of the system. The ability to extract energy from human activities has been the subject of study for many years. It's a fact: the movement of the fingers (a few mW), limb movement (about 10 mW), exhalation and inhalation, (about 100 mW), and walking (some W). The piezoelectric transducer offers a collection of higher power density than other electrostatic, especially thanks to the advantages offered by the MEMS implementation. The maximum power that is supplied to the energy harvesting devices depends on the frequency and acceleration of the vibrating system, as well as by the size of the device. A second-order model is used to analytically describe the process. From Fig. 3.2 a

Fig. 3.2 Representation of the vibration

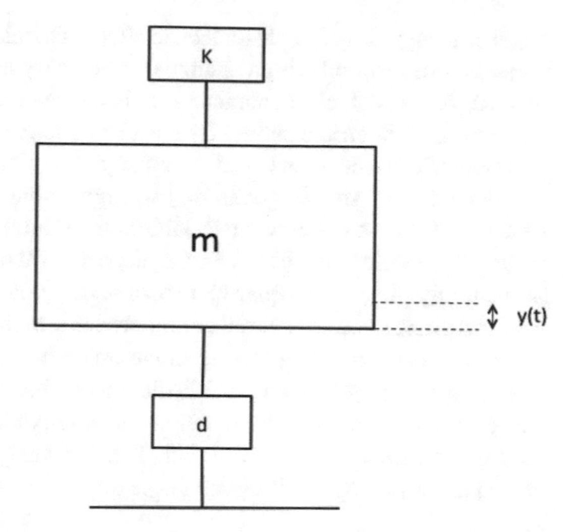

mass is suspended with a constant k connected to a rigid frame, with d is indicated the coefficient of damping. The frame is exposed in motion $x(t)$, thereby obtaining a relative movement $y(t)$ of the mass [7–10]. The maximum power occurs at the natural frequency of oscillation, namely:

$$\omega_n = \sqrt{k/n} \tag{3.2}$$

The damping coefficient can be divided into electrical and mechanical: ζ_e and ζ_p. The maximum power is generated at the frequency of resonance and can be expressed in terms of excitation X:

$$P_{\text{max}} = \frac{\zeta_e}{4(\zeta_e + \zeta_p)^2} m\omega_n^3 X^2 \tag{3.3}$$

This model shows how a highly damped system extracts energy over a wide frequency band. A less damped system would extract more power, but in a smaller frequency range.

The characterization of the vibrations can occur by means of sensors denominated accelerometers. Their advantage is to measure the absolute value and then used directly on the structure. Piezoelectric accelerometers are the most used due to the high dynamics, small size, and immunity to noise factors. Other techniques are carried out with piezoresistive or capacitive techniques that can measure static acceleration, and electrodynamic accelerometers used to measure very low frequencies. The accelerometers normally measure the linear acceleration, however, there are additional sensors which measure the acceleration in the three degrees of freedom. The block diagram of an accelerometer is displayed in Fig. 3.2, where a mass m experiences a dynamic force given by the second law of Newton, transformed into an electric signal by the piezoelectric effect. The piezoelectric effect is the ability of a material to respond with a change of electricity in response to a mechanical event (stress). A piezoelectric characteristic is the reversibility of the effect: namely, the generation of electrical energy by means of a mechanical stress, and vice versa. The piezoelectric effect is very useful in many applications are involving the production and detection of sound, generation of high voltages, microbalance, and in optical systems. There are many materials, both natural and artificial, which present a series of piezoelectric effects. Some piezoelectric materials include natural berlinite (structurally identical to quartz), brown sugar, topaz, and tourmaline. An example of artificial piezoelectric material comprises barium titanate and lead titanate zirconate. In recent years, due to growing environmental concerns regarding the toxicity in devices containing lead and RoHS directives, there has been a push to develop lead-free piezoelectric materials. To date, this initiative has led to the development of new lead-free piezoelectric materials [11–17]. The analytical model for an accelerometer can be described by the following equation:

$$y_0'' + 2\zeta\omega_o y_0' + \omega_o^2 = -g \tag{3.4}$$

where ζ is the damping coefficient linked to the inverse of the resonance frequency ω_o:

$$\omega_0 = \sqrt{k/m} \tag{3.5}$$

In particular, the piezoelectric sensors are used with high frequency sound in ultrasonic transducers for medical imaging and non-destructive industrial controls.

3.2 Thermal Energy

In any part of the earth there is a thermal gradient and then we can use this to produce energy. The thermal energy is available primarily in the industrial sector (machinery, pipes, and vehicles), in buildings, and in the human body. The temperature gradient exploits the Seebeck effect to generate electricity. To keep the thermal gradient is need of a heat source on one side and a heat sink on the other. The thermal energy is a byproduct of other forms of energy such as chemical and mechanical.

When the two ends of a conductor are at different temperatures (Fig. 3.3), a potential difference is produced between the two ends. Seebeck had thought to have created a new magnetic field mode but in reality it was an electrical voltage. The magnitude of the electromotive force V generated between the two junctions depends on the material and the temperature through the following linear relationship as a function of the Seebeck coefficient S:

$$\Delta V = S\Delta T \tag{3.6}$$

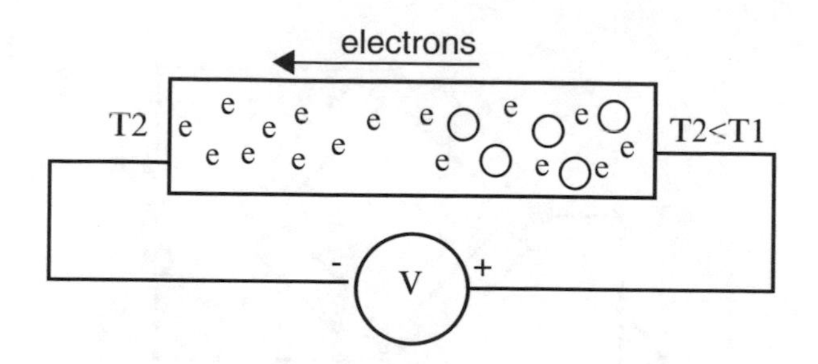

Fig. 3.3 Electronic dissemination from a cold to hot zone

Instead, the maximum power collection is given by the following relation:

$$P_e = \frac{A}{l}\left(\frac{1}{4}\frac{S^2(T_h - Tc)^2}{\rho_m}\right) \tag{3.7}$$

where A is the section of the material, ρ the resistivity of the material, l the length of the thermocouple and indicated with T are the temperatures of the hot (h) and cold (c) zone. It's necessary to measure the temperature of each side, minimizing the local voltage drop. The temperature is measured by a temperature sensor such as a thermocouple or RTD. The thermistors change the electrical resistance as function of the temperature and represent a good compromise between various sensors present on the market in terms of cost, accuracy, and response time. A thermal resistor or thermistor changes its electrical resistance with the temperature. There are thermistors with a positive temperature coefficient (PTC) and other with a negative temperature coefficient (NTC). In mathematical terms, the resistance is expressed in the following way:

$$R(T) = R_0 e^{\frac{(B(T_0 - T)}{TT_0}} \tag{3.8}$$

where $R(T)$ is the resistance at temperature T, B is a constant of the thermistor sensor, and T_0 and R_0 are, respectively, the ambient temperature of 25 °C and the corresponding resistance. The temperature coefficient is expressed in the following way:

$$\alpha = \frac{1}{R}\frac{dR}{dT} = -\frac{B}{T^2} \tag{3.9}$$

The resistance of the equation is valid in a linear range between $ln(R)$ and $1/T$, in all other intervals is used the Steinhart–Hart relation with a third-order approximation:

$$\frac{1}{T} = a + bln(R) + dln^3(R) \tag{3.10}$$

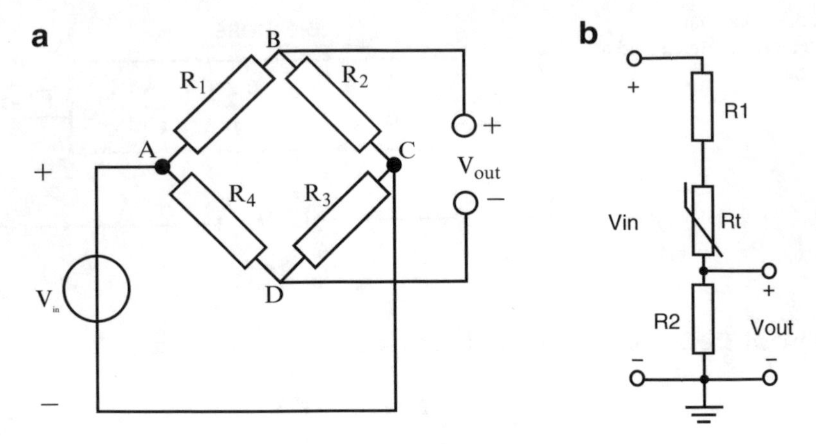

Fig. 3.4 (**a**) Wheatstone bridge and (**b**) voltage divider with the resistance of thermistor R_t

The thermistors are used in a Wheatstone bridge or voltage dividers solutions to express the electrical voltage as a function of temperature change, or variation of electrical resistance (Fig. 3.4).

In the Wheatstone bridge configuration where R_3 is the thermistor and $R_1 = R_4$ and $R_2 = R_0$, we have the following expression for the output:

$$\frac{V_{out}}{\Delta R_T} = -\frac{R_1}{(R_1 + R_0)^2} V_{in} \tag{3.11}$$

The heat generator of an assessment must be performed under certain temperature conditions, so it is necessary to establish the temperature of both sides through two thermocoolers. The corresponding temperature is controlled by means of PID adjusted to achieve a feedback and compensation control [18–23].

3.3 Electromagnetic Energy

The RF energy is currently used as cornerstones of transmitters around the world, thinking of mobile phones, Radio, base stations, TV, etc. Obviously, the energy levels are under about SAR levels. The ability to collect RF energy allows the wireless charging of low-power devices, at the same time allows to set limits in the use of the battery. The devices without battery, for example, can be designed to operate in certain time intervals or when the super-capacitor has accumulated sufficient charge to activate the electronic circuitry (Fig. 3.5).

The appeal of collecting RF energy is essentially derived from the fact that it is free energy. Over the years, the number of broadcasting stations is increased significantly, many market analysts estimate the number of mobile subscriptions

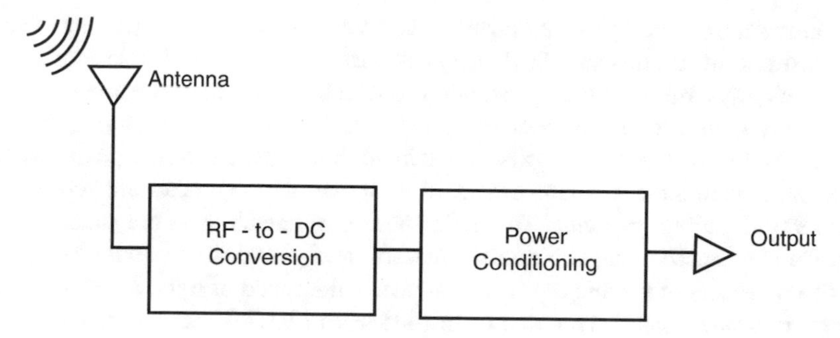

Fig. 3.5 RF energy harvesting

is growing. The mobile phones are an excellent source of RF energy and could be a source for providing energy-on-demand for a variety of applications. Thinking of how many wireless access points can find in a city center, all RF sources that we can use to recharge, for example, our smartphones. The devices such as Powerharvester receivers convert the RF energy continuously, working with standard 50 Ω antennas and provide the ability to maintain the RF-to-DC conversion efficiency in a wide range of operating conditions, including variations of power in input and output load resistance. The RF energy can be used to charge a wide range of low-power devices. At close range, we can recharge GPS or wearable medical sensors and a wide range of consumer electronics. Depending on the required power, the power can be sent in a continuous mode, on a scheduled basis, or on-demand [24–30].

3.4 Space Radiation

The radiation is a form of energy composed of high speed particles. Typically there are two types of radiation: ionizing and non-ionizing. The first they have a lot of energy, enough to interfere with the atom and the electron to modify the state, unlike the other radiation.

Ionizing radiation includes gamma rays, *x*-rays, protons, electrons, neutrons, alpha and beta particles; and of non-ionizing radiation includes microwaves, visible light, infrared, and radio frequency waves. The spatial radiation is ionizing and consists of highly energetic charged particles.

The solar wind is a stream of charged particles that originated in the upper layers of the sun. It consists mainly of electrons and protons and blows constantly from the surface of the sun. The energy possessed by these charged particles is between 1.5 and 10 KeV. The average speed of these particles is about 145 km/s. This speed is lower than the solar escape velocity of 618 km/s. However, some of the particles are able to have sufficient energy to reach the terminal velocity of 400 km/s. So, they are allowed to create the solar wind. At the same temperature, the electrons reach escape velocity because of their smaller mass which helps to build up an electric field that accelerates the charged particles further. The solar wind can travel up to the distance of 75 astronomical units and the density can vary from 1 to 10 particles/cm^3.

Cosmic rays are high-energy radiation to which the origin of the universe. They are mainly composed of high-energy protons and atomic nuclei ranging from the lightest to the heaviest. They also contain high-energy electrons, positrons, and other subatomic such as muons. The 90 % of cosmic ray nuclei are protons and about 9 % are alpha particles. The main sources of cosmic rays are supernovae of massive stars, active galactic nuclei, quasars, and gamma-ray bursts. The highly charged particles of cosmic rays travel at nearly the speed of light. Most of galactic cosmic rays have energies ranging from 100 MeV to 10 GeV. Since, cosmic rays are electrically charged, they are deflected by magnetic fields, and their directions are random without giving the exact idea of their origin [31–35].

3.5 Solar Radiation

The Sun is a star classified as "yellow dwarf" (surface temperature of about 5570 °C) consists primarily of hydrogen (74 % of the mass) and helium (24–25 % of the mass). The energy radiated by the sun derives from a nuclear fusion reaction: every second approximately $6.2 * 10^{11}$ kg of hydrogen is transformed into helium, with a mass loss of about 4.26103 kg; these are transformed into energy according to Einstein's relation $E = mc^2$, resulting in an energy of $3846 * 10^{26}$ J, which corresponds to a radiated power $P_{tot} = 3846 * 10^{26}$ W. It's possible to approximate the sun to a point source placed at a distance from the earth equal to the average distance during the year ($D = 149 * 10^6$ km) that radiates energy in a manner uniformly distributed throughout the solid angle (Fig. 3.6). Consequently, the energy flow is constant in the unit of surface of radius D and its value is equal to one divided by the entire surface of the sphere.

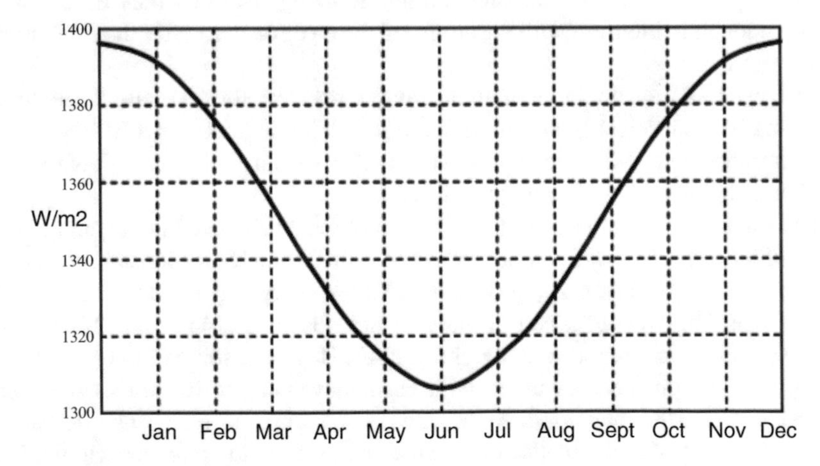

Fig. 3.6 Mean solar radiation during 1 year

The power of the incident solar radiation on a surface is called the solar radiation and is measured in W/m^2. Some considerations about the solar radiation:

- It may be seem surprising that from such big numbers we've got to calculate a value for the power available per square meter outside the atmosphere so close to the typical values cited in the literature;
- Also, how is it possible to know with such precision the power produced in the Sun?
- Actually it proceeds in exactly the opposite: it measures the power per square meter available outside of the atmosphere and, by reversing the calculation, it is estimated the power radiated by the sun.

The energy received on Earth from the Sun in a year is about 10,000 times the current energy needs. If we consider a conversion efficiency of 10 % is that it would require an area of solar panels equal to the area of England, that is one-thousandth of the Earth's surface exposed to the Sun (England area = 130.325 km^2, the Earth's surface exposed to the Sun $= 127.796 * 10^6\ km^2$, and the ratio is 1/980.6).

Conventionally, we define the constant solar radiation outside the atmosphere equal to 1360 W/m^2. It is interesting to analyze the spectrum of the radiation emitted by the Sun shown in Fig. 3.7. In particular, some considerations are reported below:

- The specter measured outside the atmosphere corresponds with high precision to the theoretical emitted by a black body whose surface temperature is that of the sun.

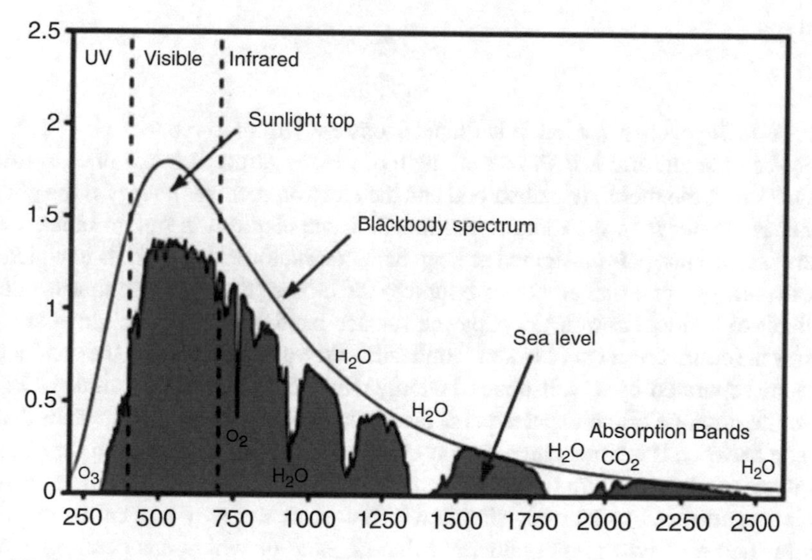

Fig. 3.7 Spectral irradiance $W/M^2/nm$ as function of wavelength (nm)

- The bulk of the emitted radiation is concentrated in the visible wave lengths (approximately between 400 and 700 nm).
- The wavelength of the radiation is very important for the interaction with the photovoltaic cell.
- The spectrum at sea level is significantly different: in general it is attenuated, but mainly much more jagged and full of "holes": these are due to the fact that certain atomic species and the water present in the atmosphere completely absorb the radiation of specific wavelengths.

As for the spectrum and the power available at sea level they vary depending on the atmospheric conditions and of the atmosphere traversed thickness, which in turn depends on the position of the sun in the sky. The minimum thickness is at its zenith, that is when the Sun occupies a position vertically above the observer's head [30–35].

3.5.1 Photovoltaic Cell

The solar cells exploit the photoelectric effect, i.e., the effect for which a suitable material, for example, a metal, emits electrons by light radiation. The mechanism can be explained by assuming to represent the light like a beam of particles called photons (alternate interpretation of the light wave phenomena).

Each photon, characterized by a certain wavelength, carries a well-defined amount of energy according to the relation:

$$E = \frac{hc}{\lambda} \tag{3.12}$$

where E is the energy carried, h is Plank's constant (equal to $6.626 * 10^{-34}$ J s), c is the speed of light, and λ is the wavelength of the radiation. When a photon strikes an electron of the metal, it is absorbed and the electron receives energy if the photon energy is greater than the metal work function: the electron is free to break away; otherwise the energy is dissipated as heat. Semiconductors are materials in which the vast majority of external electrons bound to the atoms (that is, those that would be available to conduct current) occupy the valence band, and only a minimal amount occupy the conduction band. In semiconductors the valence band and the conduction band are separated by a well-defined energy gap, for silicon, for example, 1.12 eV. If we illuminate a silicon surface and the photons have energy higher than that of the gap between the bands, an electron of the valence band can absorb the photon, and it can work to acquire the energy and move to the conduction band to conduct an electric current. A favorable situation is in a particular structure called junction. It is the union of two semiconductor volumes, each of which has been separately worked by introducing the impurities in the interior (doping). For example, silicon is a tetravalent element (each atom is bonded to four other atoms equal, by means of the four valence electrons available external orbital). Adding small amounts of a pentavalent element (such as phosphorus or arsenic), the fifth electron not engaged in a link goes to occupy the conduction band, by resulting in an excess of free negative carriers. There is talk of doping n. Adding small amounts of tetravalent elements (such as boron), instead, they are generated of unstable bonds with the surrounding atoms that tend to trap an electron to stabilize, by leaving a positive charge is not offset gap call, and then an excess of positive charges available to conduct current. It is important to stress that although there is an excess of positive and negative charges available to run, a total of the two volumes of doped material are electrically neutral (the total number of positive charges inside them exactly compensates the number of negative charges). When the two volumes of the excess electrons of the n tends zone are contacted to diffuse in the p region; conversely the excess of p tends to spread in the gaps area n area. This migration produces an imbalance of electric charge that determines the birth of an electric field that opposes the diffusion and attracts electrons to zone n and p gaps towards the area. The equilibrium is reached when the number of charges that moves in one direction due to diffusion is offset by an equal number of charges that moves in the opposite direction due to the electric field (drift, Fig. 3.8).

In the single cell, the photocurrent generation coexists with the mechanisms that regulate the normal flow of current in a p–n junction to vary the voltage at its ends. The photocurrent has an intensity that depends on many factors: radiation, angle of incidence, temperature, type of semiconductor, etc. The ideal model is shown in Fig. 3.9.

The I_D has the classic expression of the current of a diode:

$$I_D = I_S(e^{\frac{qV_C}{nKT}} - 1) \tag{3.13}$$

where q is the electron charge, k is the Boltzmann constant, T is the temperature in Kelvin, n is the ideality factor or emission coefficient (between 1 and 2, by depending on the manufacturing process), and I_S is the saturation current: the value depends on the characteristics of the diode. The current delivered by the cell is given by:

$$I_C = I_L - I_S(e^{\frac{qV_C}{nKT}} - 1) \tag{3.14}$$

The typical load output voltage of a cell is of the order of 0.5–0.6 V (not in case very close to the classical voltage of a diode in conduction). Typical values of photocurrent for a cell in silicon are of the order of 30 mA/cm^2. One of the most popular formats for the cells is approximately square in shape and 125 mm side, for a total surface of about 156 cm^2; the short-circuit current is therefore at around 4 A. For the intermediate work points between the open circuit and short circuit the characteristic of the cell varies according to a graph shown in Fig. 3.10.

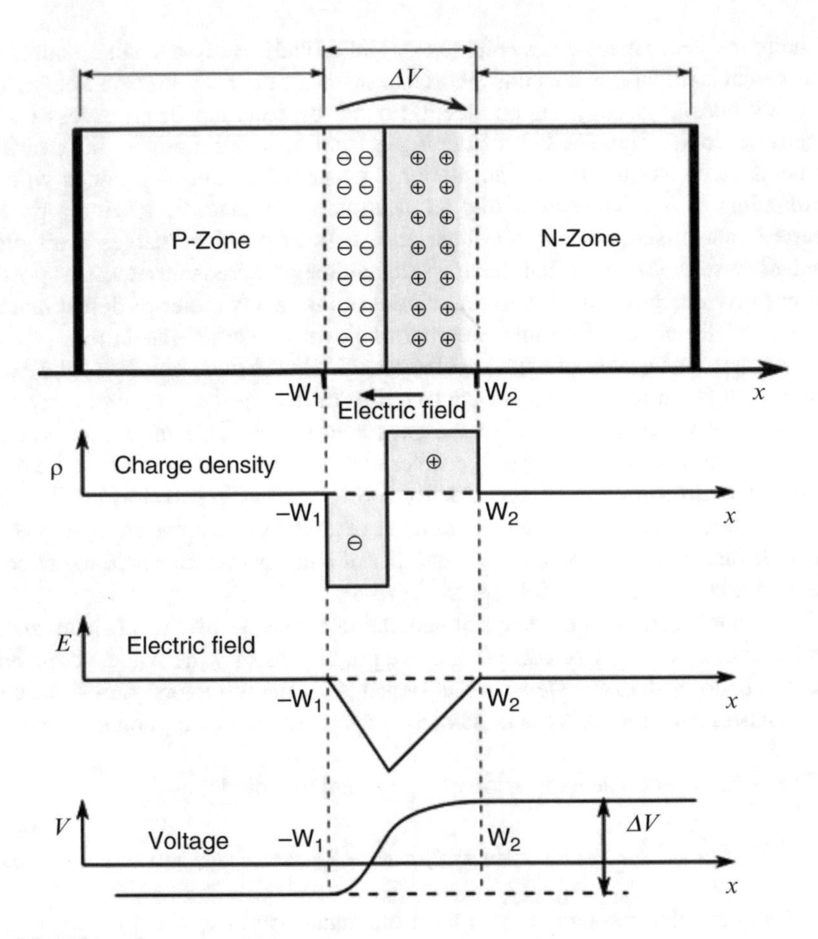

Fig. 3.8 PN junction

The output power curve shown in Fig. 3.10 has a maximum (MPP—Maximum Power Point) which correspond to the values of current and voltage IMPP and VMPP. The Maximum Power Point (MPP) changes to vary of the characteristics of the cell. The manufacturers characterize the cells through the CT temperature coefficient that indicates the variation of delivered power as function of the temperature. Typical values are between −0.2 and −0.5 %. The efficiency of the photovoltaic cell is the ratio of the electric power output and the power incident on the cell. The conversion efficiency varies greatly at different technology used, but in general is rather low by ranging from 5 to 8 % for the cells of amorphous silicon to 20 % for multi-junction cells.

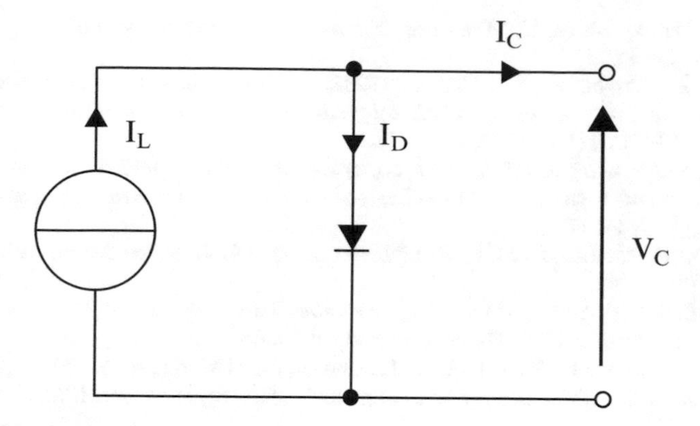

Fig. 3.9 Model of a photovoltaic cell

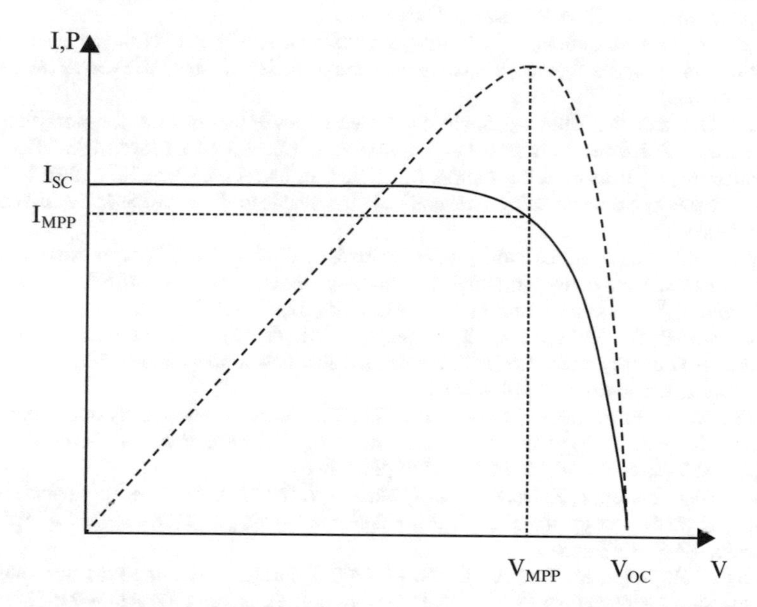

Fig. 3.10 I/V characteristic of a photovoltaic cell and output power (*dashed line*)

References

1. Park, J., & Mackqy, S. (2003). *Practical Data Acquisition for instrumentation and system control*. Elsevier, Oxford.
2. Lacanette, K. (2003). *National temperature sensors handbook*. Annali di Matematica Pura ed Applicata. National Semiconductor.
3. National Instruments. (1996). *Data acquisition fundamentals, application note 007.*
4. National Instruments. (1996). *Signal conditioning fundamentals for PC-based data acquisition*, Application Note 048
5. Taylor, J. (1986). *Computer-based data acquisition system*. Instrument Society of America.

6. Di Paolo Emilio, M. (2013). *Data acquisition system, from fundamentals to applied design*. New York: Springer.
7. Roundy, S., Wright, P., & Pister. K. (2002). Micro-electrostatic vibration-to- electricity converters. In *Proceedings of ASME international mechanical engineering congress and exposition (IMECE)* (Vol. 220, pp. 17–22).
8. Stordeur, M., & Stark, I. (1997). Low power thermoelectric generator: self-sufficient energy supply for micro systems. In *Proceedings of the 16th international conference on thermoelectrics* (pp. 575–577).
9. Shenck, N., & Paradiso, J. (2001). Energy scavenging with shoe-mounted piezoelectrics. *Micro IEEE, 21*(3), 30–42.
10. Roundy, S. (2003). *Energy scavenging for wireless sensor nodes with a focus on vibration to electricity conversion*. Ph.D Thesis, University of California.
11. Tsutsumino, T., Suzuki, Y., Kasagi, N., Kashiwagi, K., & Morizawa, Y. (2006). *Micro seismic electret generator for energy harvesting*. Technical Digest PowerMEMS (pp. 133–136). Berkeley, USA.
12. Sterken, T., Altena, G., Fiorini, P., & Puers, R. (2007). *Characterisation of an electrostatic vibration harvester*, EDA Publishing Association.
13. Sterken, T., Baert, K., Puers, R., & Borghs, S. (2002). Power extraction from ambient vibration. In *Proceedings of the SeSens (Workshop on Semiconductor Sensors, Veldhoven, Netherlands)* (pp. 680–683).
14. Szarka, G., Stark, B., & Burrow, S. (2012). Review of power management for energy harvesting systems. *IEEE Transactions on Power Electronics, 27*(2), 803–815. ISSN: 0885-8993.
15. Cammarano, A., Burrow, S. G., Barton, D. A. W., Carrella, A., & Clare, L. R. (2010). Tuning a resonant energy harvester using a generalized electrical load. *Smart Materials and Structures, 19*, 055003.
16. Guyomar, D., Badel, A., Lefeuvre, E., & Richard, C. (2005). Toward energy harvesting using active materials and conversion improvement by nonlinear processing. *IEEE Transactions on Ultrasonics, Ferroelectrics, and Frequency Control, 52*, 584–595.
17. Mitcheson, P. D., Stoianov, I., & Yeatman, E. M. (2012). Power-extraction circuits for piezoelectric energy harvesters in miniature and low-power applications. *IEEE Transactions on Power Electronics, 27*, 4514–4529.
18. Szarka, G. D., Burrow, S. G., & Stark, B.H. (2012). Ultra-low power, fully-autonomous boost rectifier for electro-magnetic energy harvesters. *IEEE Transactions on Power Electronics, 28*(7), 3353–3362. doi:10.1109/TPEL.2012.2219594.
19. Maurath, D., Becker, P. F., Spreeman, D., Manoli, Y. (2012). Efficient energy harvesting with electromagnetic energy transducers using active low-voltage. *IEEE Journal of Solid-State Circuits, 47*(6), 1369–1380
20. Beeby, S. P., Tudor, M. J., & White, N. M. (2006). Energy harvesting vibration sources for microsystems applications. *Measurement Science and Technology, 17*, R175–R195.
21. Khaligh, A., Zeng, P., & Zheng, C. (2010). Kinetic energy harvesting using piezoelectric and electromagnetic technologies—state of the art. *IEEE Transactions on Industrial Electronics, 57*(3), 850–860.
22. Paulo, J., & Gaspar, P. D. (2010). Review and future trend of energy harvesting methods for portable medical devices. In *Proceedings of the World Congress on Engineering* (Vol. 2)
23. Zhu, D., Tudor, M. J., Beeby, S. P. (2010). Strategies for increasing the operating frequency range of vibration energy harvesters: A review. *Measurement Science and Technology, 21*, 022001-1–022001-29.
24. Cepnik, C., Lausecker, R., & Wallrabe, U. (2013). Review on electrodynamic energy harvesters—a classification approach. *Micromachines, 4*(2), 168–196. http://www.mdpi.com/2072-666X/4/2/168. Accessed 20 Jan 2015.
25. Ulaby, F. T., Michielssen, E., & Ravaioli, U. (2010). *Fundamentals of Applied Electromagnetics* (6th ed.). Prentice Hall, USA.
26. Roundy, S., Wright, P. K., & Rabaey, J. M. (2003). A study of low level vibrations as a power source for wireless sensor nodes. *Computer Communications, 26*(11), 1131–1144.

27. Sazonov, E., Li, H., Curry, D., & Pillay, P. (2009). Self-powered sensors for monitoring of highway bridges. *IEEE Sensors Journal, 9*, 1422–1429.
28. Toh, T. T., Mitcheson, P. D., Holmes, A. S., & Yeatman, E. M. (2008). A continuously rotating energy harvester with maximum power point tracking. *Journal of Micromechanics and Microengineering, 18*, 104008-1-7.
29. Howey, D. A., Bansal, A., & Holmes, A. S. (2011). Design and performance of a centimetre-scale shrouded wind turbine for energy harvesting. *Smart Materials and Structures, 20*, 085021.
30. Razavi, B. (2002). *Design of analog CMOS integrated circuits*. McGraw-Hill
31. Razavi, B. (2008). *Fundamentals of microelectronics*. New York: Wiley.
32. Sedra, A. S., & Smith, K. C. (2013). *Microelectronic circuits*. Oxford: Oxford University.
33. Razavi, B. (2002). *Design of integrated circuits for optical communications*. McGraw-Hill.
34. Hurst, P. J. (2001). *Analysis and design of analog integrated circuits*. New York: Wiley.
35. Spies, P. (2015). *Handbook of energy harvesting power supplies and applications*. CRC Press Book, France.

Chapter 4
Electromagnetic Transducers

4.1 Introduction

The study of RF signals implies a clear distinction between RF nonradiative and radiative. The first is based on an inductive coupling, while the second uses the transmission and reception of radio waves. The harvesting energy is a process where the energy from the environment goes into a load. With the RF energy transfer, however, it is the process that uses a dedicated RF source for the power wirelessly. The transmission of RF signals transmitted in the collection process is involuntary arising from various systems such as smartphones. The transfer of RF energy is mostly used in inductive systems, according to the Wi-Fi standards. In these cases two coils are placed in proximity to obtain RF transference, two coils in the vicinity form an electrical transformer. In the circuit of Fig. 4.1 is displayed a general scheme where two coils, one in reception and transmission in the other, form the transfer system by using a capacitor to rectify the wave shape [1–10].

RF energy transfer over a distance is need to use the radiative transfer: an RF source (not intentional) is connected to an antenna that emits radio waves. At a certain distance, a receiving antenna picks up part of the converting waves into an electrical signal and transferred to a load (Fig. 4.2). The power transfer between the transmitting and receiving antenna is described by Friis equation.

4.2 Electromagnetic Waves and Antenna

The electromagnetic waves can travel in space or in a dielectric. The sound waves are examples of mechanical waves, in contrast to the bright ones (photons) that represent a type of electromagnetic waves. The electromagnetic waves are created by the vibration of an electric charge: a wave with an electric and magnetic component is created. The energy transported has a velocity in vacuum equal to that

M. Di Paolo Emilio, *Microelectronic Circuit Design for Energy Harvesting Systems*, DOI 10.1007/978-3-319-47587-5_4

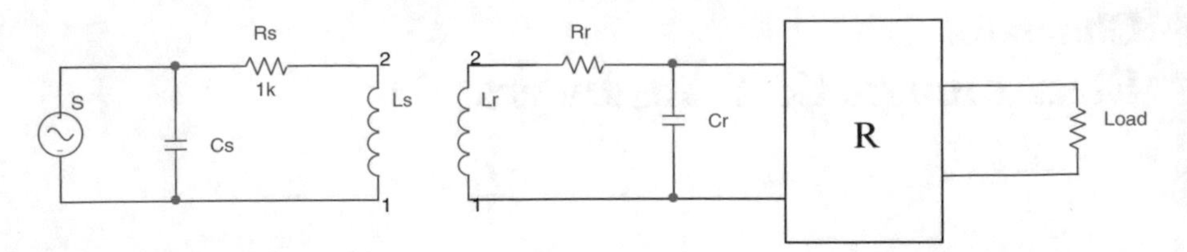

Fig. 4.1 General layout of an RF energy harvesting. The R block is used to rectify the signal

Fig. 4.2 Diagram block for RF energy harvesting

of light indicated with the letters c, in a medium speed it is less than c. The transport mechanism exploits the phenomenon of absorption and re-emission of the wave energy from the atoms of the material. When an electromagnetic wave affects atoms, the energy is absorbed with vibration of the electrons that make up the material. After a short period of vibrational motion, the vibrating electrons create a new electromagnetic wave with the same frequency of the first electromagnetic wave. These vibrations occur only for a very short time and the wave motion is delayed through the medium. Once the electromagnetic wave energy is re-emitted by an atom, it travels through a small region of space between atoms. Once the next atom reached, the electromagnetic wave is absorbed, transformed into electron vibration, and then re-emitted as an electromagnetic wave. As part of the energy harvesting, the Friis's equation used in telecommunications engineering provides us the transmitted power of an antenna in ideal conditions to another antenna at a certain distance. An antenna is an electrical device that converts electrical energy into radio waves, and vice versa. It is usually used with a radio transmitter or radio receiver. In the transmission, a radio transmitter provides a radio frequency oscillating electric current (that is, a high frequency alternating current (AC)) to the antenna that radiates the energy from the current as electromagnetic waves (radio waves). In reception, an antenna intercepts a part of the power of an electromagnetic wave to produce a small voltage at its terminals where is applied a receiver to amplify the signal. The electromagnetic waves are described by Maxwell's equations which represent one of the most elegant and concise ways to affirm the fundamentals of electricity and magnetism. From them have been developed most of the industry working relationships [11–20]. The four equations of Maxwell describe the electric and magnetic fields arising from the distribution of electrical charges, and how they change over time. The formulation of the equations is a study of many scientists and deep insight of Michael Faraday. The first equation expresses the Gauss' law for electric fields: the integral of the electric field output on a surface that encloses a volume is equal to the internal total charge:

$$\int \vec{E} d\vec{A} = \frac{q}{\epsilon_0} \tag{4.1}$$

The second equation is analogous to the magnetic fields:

$$\int \vec{B} d\vec{A} = 0 \tag{4.2}$$

The first two equations of Maxwell are integrals of the electric and magnetic fields on closed surfaces. The other two equations of Maxwell, discussed below, are integral of electric and magnetic fields around closed curves which represent the work required to take a charge around a closed curve in an electric field, and the similar in the magnetic field. The third is the Faraday's law of induction and the fourth is the Ampere's law:

$$\oint_C \vec{E} d\vec{l} = -\frac{d}{dt} \int \vec{B} d\vec{A} \tag{4.3}$$

The first term is integrated around a closed line, usually a wire, and provides the total variation of the voltage on the circuit that is generated by a variable magnetic field.

$$\oint_C \vec{B} d\vec{l} = \mu_0 (I + \frac{d}{dt} \epsilon_0 \int \vec{E} d\vec{A}) \tag{4.4}$$

In differential form the equations for the electromagnetic waves become the following:

$$\nabla \cdot E = 4\pi\rho \tag{4.5}$$

$$\nabla x E = -\frac{1}{c} \frac{\partial B}{\partial t} \tag{4.6}$$

$$\nabla \cdot B = 0 \tag{4.7}$$

$$\nabla x B = \frac{4\pi}{c} J + \frac{1}{c} \frac{\partial E}{\partial t} \tag{4.8}$$

where E is the electric field, B the magnetic field, J the current density, and ρ the charge density. After the definition of the electromagnetic waves, now we can evaluate the engineering or how to transmit them by means of antennas, and then capture them to produce electric current. To begin with the derivation of the Friis equation considering two antennas in free space (obstructions nearby) separated by a distance R (Fig. 4.3).

Now, we suppose a transmission antenna omnidirectional, without loss, and the receiving antenna is far from the transmission range. P_t is the transmission power and p the power density (in watts per square meter) of the wave received on the antenna at a distance R from the transmission antenna:

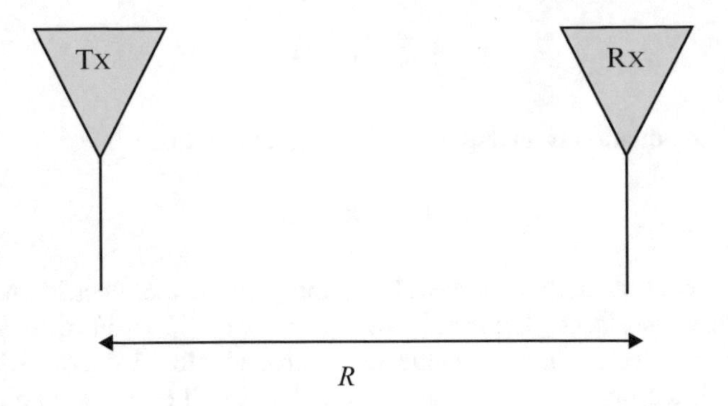

Fig. 4.3 Transmission and reception layout

$$p = \frac{P_T}{4\pi R^2} G_t \tag{4.9}$$

where the parameter G_t is the antenna gain. If the receiving antenna has an effective capture area A_{er}, then the power received by the antenna is the following:

$$P_r = \frac{P_T}{4\pi R^2} G_t A_{\text{er}} \tag{4.10}$$

where the effective capture area can be expressed in the following mode:

$$A_{\text{er}} = \frac{\lambda}{4\pi} G \tag{4.11}$$

And then the Friis's formula, namely the expression for the power P_r can be expressed as follows:

$$P_r = \frac{P_T G_t G_r}{(4\pi R)^2} \lambda^2 \tag{4.12}$$

To decide on the feasibility of RF energy harvesting, we need to assess the power levels. Based on the state-of-the-art power consumption of sensors available on the market, our goal is to power the level of about 100 μW. For RF energy harvesting, the most interesting systems to be explored in Europe are GSM900 (downlink: 935–960 MHz), GSM1800 (downlink: 1805.2–1879.8 MHz), and Wi-Fi World (2.4–2.5 GHz). All these systems guarantee an excellent presence and size of antennas very small about of 10–50 cm^2. Many studies have estimated power density in the GSM900 band equal to a 0.01–0.3 μWcm^{-2} range between 25 and 100 m from a base station. This implies dimensions of antennas of the order of 300–1000 cm^2 to achieve power levels of 100 uW. The technological limitations are related to the maximum power allowed to transmit and on the limitation of transmission between transmitter power and antenna gain: their product is known as EIRP, effective isotropic power [20–35].

4.3 System Design

The RF energy harvesting focuses on the receiving layout that needs some tricks to be able to feed the load properly. The block diagram of receiving part is shown in Fig. 4.4.

The general layout is composed of a receiving antenna, a rectifier connected to an RF-side to the receiving antenna and on the DC side of a load. In general, an impedance matching circuit is applied between the antenna and the rectifier. In general there is a DC-DC conversion circuit connected to an energy storage system (battery or capacitor) which is connected to the load. Considering the equivalent antenna to a voltage source in series with a resistor, the equivalent circuit is described in Fig. 4.5 where the load and the conversion part is combined in a load single Y_L. The matching network is presented as an L network.

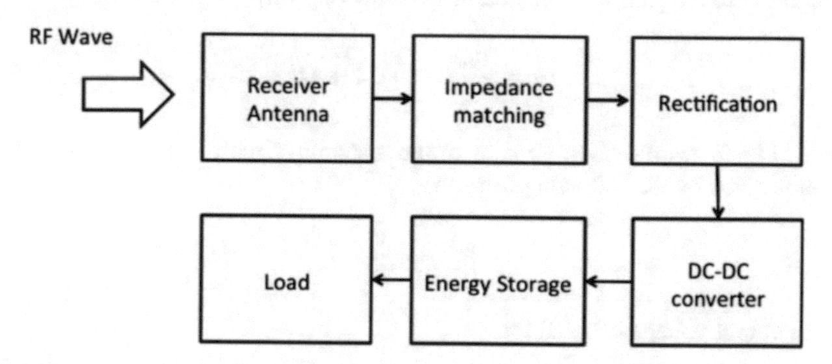

Fig. 4.4 Block diagram of receiving system

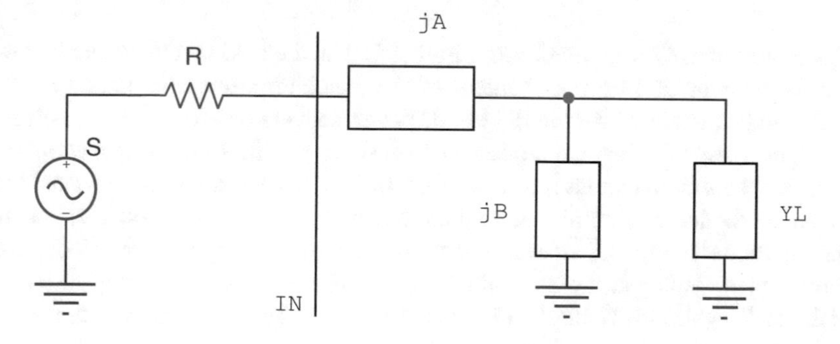

Fig. 4.5 Antenna equivalent circuit and L matching network

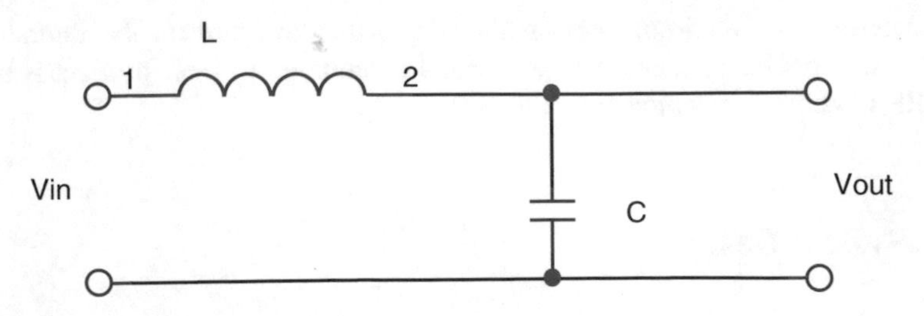

Fig. 4.6 LC voltage boosting circuit

The input power is given by the following expression:

$$P_{\text{in}} = \frac{1}{2} v_{\text{in}}^2 Re(Y_{\text{in}}) \tag{4.13}$$

where Y is the admittance, i.e., the inverse of the impedance. In output, instead, the power is given by the same expression calculated for Y_L. To connect the rectifier to the antenna has been proposed a voltage boosting circuit: a classic example is the LC circuit (Fig. 4.6).

In this case the gain is expressed in the following way:

$$g = \frac{v_{\text{out}}}{v_s} = \frac{1}{2}\sqrt{1 + Q^2} \tag{4.14}$$

where Q indicates the quality factor of the matching circuit. Whereas the incident power is given by the following formula:

$$P_{\text{inc}} = \frac{v_s}{8R^2} \tag{4.15}$$

The output voltage is equal to:

$$v_{\text{out}} = \frac{1}{2}\sqrt{(1 + Q^2) * 8R^2 P_{\text{inc}}} \tag{4.16}$$

The input voltage of the rectifier/multiplier is then dictated by the available power from the antenna and by the Q factor of the matching circuit. To maximize v_{out}, which will be beneficial for the RF-to-DC power conversion efficiency, as well as to the output voltage DC level, there is need to design a rectifier/multiplier having a real part of input admittance that is as low as possible. The easiest rectifier is constituted by a single diode. In general, due to the rapid switching speed, it considers using a Schottky diode. In Fig. 4.7 an example of circuit with Schottky rectifier diode; in the circuit is also displayed the equivalent rectifier circuit. The source has an internal resistance R_g while the Schottky diode is represented by a resistance that expresses

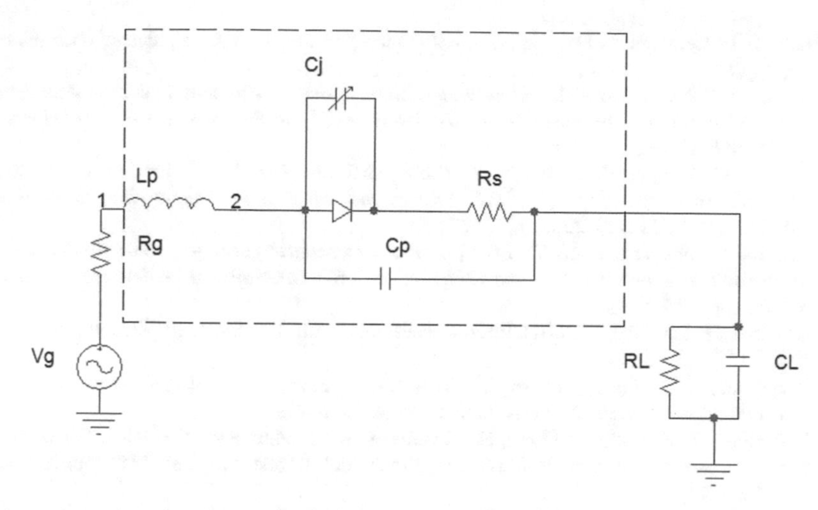

Fig. 4.7 Equivalent rectifier circuit

the conduction losses in the substrate, while C_j is the parasitic capacitance expresses by the following relation of proportionality as function of the frequency:

$$C_j(f) \propto \frac{1}{\sqrt{1 - \frac{f}{\phi}}} \tag{4.17}$$

where ϕ is the potential barrier. All these parameters can be determined by using the data sheet of the single component. The antenna must be connected to a rectifier/multiplier. Once the frequency band or frequency has been decided, there is need to start determining the impedance of the input rectifier/multiplier. If we want to design an antenna without by using an impedance matching circuit, it is necessary to design an antenna input impedance equals to the complex conjugate of the impedance of the rectifier input. In addition, we have to design an antenna with a low resistive part of the input impedance. To achieve this antenna, we must make the antenna electrically small. In addition, we have to design the antenna that has a relatively high inductance, which means that we need a small loop antenna. The RF energy harvesting systems use a recharge battery and an array of antennas to provide a sufficient energy source to the load.

References

1. Park, J., & Mackqy, S. (2003). *Practical Data Acquisition for instrumentation and system control*. Elsevier, Oxford.
2. Lacanette, K. (2003). *National temperature sensors handbook*. Annali di Matematica Pura ed Applicata. National Semiconductor.
3. National Instruments. (1996). *Data acquisition fundamentals, Application note 007*.

4. National Instruments. (1996). *Signal conditioning fundamentals for PC-based data acquisition, Application Note 048.*
5. Taylor, J. (1986). *Computer-based data acquisition system.* Instrument Society of America.
6. Di Paolo Emilio, M. (2013). *Data acquisition system, from fundamentals to applied design.* New York: Springer.
7. Roundy, S., Wright, P., & Pister. K. (2002). Micro-electrostatic vibration-to- electricity converters. In *Proceedings of ASME international mechanical engineering congress and exposition (IMECE)* (Vol. 220, pp. 17–22).
8. Stordeur, M., & Stark, I. (1997). Low power thermoelectric generator: self-sufficient energy supply for micro systems. In *Proceedings of the 16th international conference on thermoelectrics* (pp. 575–577).
9. Shenck, N., & Paradiso, J. (2001). Energy scavenging with shoe-mounted piezoelectrics. *Micro IEEE, 21*(3), 30–42.
10. Roundy, S. (2003). *Energy scavenging for wireless sensor nodes with a focus on vibration to electricity conversion*. Ph.D Thesis, University of California.
11. Tsutsumino, T., Suzuki, Y., Kasagi, N., Kashiwagi, K., & Morizawa, Y. (2006). *Micro seismic electret generator for energy harvesting*. Technical Digest PowerMEMS (pp. 133–136). Berkeley, USA.
12. Sterken, T., Altena, G., Fiorini, P., & Puers, R. (2007). *Characterisation of an electrostatic vibration harvester, EDA Publishing Association.*
13. Sterken, T., Baert, K., Puers, R., & Borghs, S. (2002). Power extraction from ambient vibration. In *Proceedings of the SeSens (Workshop on Semiconductor Sensors, Veldhoven, Netherlands)* (pp. 680–683).
14. Szarka, G., Stark, B., & Burrow, S. (2012). Review of power management for energy harvesting systems. *IEEE Transactions on Power Electronics, 27*(2), 803–815. ISSN: 0885-8993.
15. Cammarano, A., Burrow, S. G., Barton, D. A. W., Carrella, A., & Clare, L. R. (2010). Tuning a resonant energy harvester using a generalized electrical load. *Smart Materials and Structures, 19*, 055003.
16. Guyomar, D., Badel, A., Lefeuvre, E., & Richard, C. (2005). Toward energy harvesting using active materials and conversion improvement by nonlinear processing. *IEEE Transactions on Ultrasonics, Ferroelectrics, and Frequency Control, 52*, 584–595.
17. Mitcheson, P. D., Stoianov, I., & Yeatman, E. M. (2012). Power-extraction circuits for piezoelectric energy harvesters in miniature and low-power applications. *IEEE Transactions on Power Electronics, 27*, 4514–4529.
18. Szarka, G. D., Burrow, S. G., & Stark, B.H. (2012). Ultra-low power, fully-autonomous boost rectifier for electro-magnetic energy harvesters. *IEEE Transactions on Power Electronics, 28*(7), 3353–3362. doi:10.1109/TPEL.2012.2219594.
19. Maurath, D., Becker, P. F., Spreeman, D., & Manoli, Y. (2012). Efficient energy harvesting with electromagnetic energy transducers using active low-voltage. *IEEE Journal of Solid-State Circuits, 47*(6), 1369–1380.
20. Beeby, S. P., Tudor, M. J., & White, N. M. (2006). Energy harvesting vibration sources for microsystems applications. *Measurement Science and Technology, 17*, R175–R195.
21. Khaligh, A., Zeng, P., & Zheng, C. (2010). Kinetic energy harvesting using piezoelectric and electromagnetic technologies—state of the art. *IEEE Transactions on Industrial Electronics, 57*(3), 850–860.
22. Paulo, J., & Gaspar, P. D. (2010). Review and future trend of energy harvesting methods for portable medical devices. In *Proceedings of the World Congress on Engineering* (Vol. 2)
23. Zhu, D., Tudor, M. J., Beeby, S. P. (2010). Strategies for increasing the operating frequency range of vibration energy harvesters: A review. *Measurement Science and Technology, 21*, 022001-1–022001-29.
24. Cepnik, C., Lausecker, R., & Wallrabe, U. (2013). Review on electrodynamic energy harvesters—a classification approach. *Micromachines, 4*(2), 168–196. http://www.mdpi.com/2072-666X/4/2/168. Accessed 20 Jan 2015.

25. Ulaby, F. T., Michielssen, E., & Ravaioli, U. (2010). *Fundamentals of Applied Electromagnetics* (6th ed.). Prentice Hall, USA.
26. Roundy, S., Wright, P. K., & Rabaey, J. M. (2003). A study of low level vibrations as a power source for wireless sensor nodes. *Computer Communications, 26*(11), 1131–1144.
27. Sazonov, E., Li, H., Curry, D., & Pillay, P. (2009). Self-powered sensors for monitoring of highway bridges. *IEEE Sensors Journal, 9*, 1422–1429.
28. Toh, T. T., Mitcheson, P. D., Holmes, A. S., & Yeatman, E. M. (2008). A continuously rotating energy harvester with maximum power point tracking. *Journal of Micromechanics and Microengineering, 18*, 104008-1-7.
29. Howey, D. A., Bansal, A., & Holmes, A. S. (2011). Design and performance of a centimetre-scale shrouded wind turbine for energy harvesting. *Smart Materials and Structures, 20*, 085021.
30. Razavi, B. (2002). *Design of analog CMOS integrated circuits*. McGraw-Hill
31. Razavi, B. (2008). *Fundamentals of microelectronics*. New York: Wiley.
32. Sedra, A. S., & Smith, K. C. (2013). *Microelectronic circuits*. Oxford: Oxford University.
33. Razavi, B. (2002). *Design of integrated circuits for optical communications*. McGraw-Hill.
34. Hurst, P. J. (2001). *Analysis and design of analog integrated circuits*. New York: Wiley.
35. Spies, P. (2015). *Handbook of energy harvesting power supplies and applications*. CRC Press Book, France.

Chapter 5
Piezoelectric Transducers

5.1 Introduction

The piezoelectric effect had its origin in the late 1800s, where the French Curie discovered some crystals polarization effects subjected to mechanical stress. Applying an electric field there was also a deformation with implications in the field of telecommunications. With the passage of time there were many solutions in lead zirconate and materials PZT (zirconate-titanate). These latter have become the dominant materials for a variety of applications such as actuators and ultrasonic medical devices. Subsequently, also polymers such as polyvinylidene difluoride (PVDF) have been found to have piezoelectric properties because of the stretching of the molecules. Today, the most important applications of piezoceramics are in medicine as ultrasonic devices, in the measurements of time (quartz) and the fuel technology. New applications in micro-energy harvesting need further technical and economic progress [1–15].

5.2 Materials

The single crystals such as quartz, $LiNbO_3$, $GaPO_4$, or Langasite ($La_3Ga_5SiO_{14}$) are less commonly used as piezoelectric devices with respect to Pb (ZRX Ti1-x) O_3 (PZT), but there are, however, some commonly used applications involving high frequencies or that require resistance to high temperatures. The process for producing polycrystalline piezoceramic materials generally comprises two stages. After preparation of the ceramic powder, it provides for the cooking of a mixture of the oxide powder (calcination) and then milling into fine powder, the ceramic is sintered to the desired shape. In the preparation of Pb (ZRX Ti1-x) O_3 the oxide powders of PbO, ZrO_2, and TiO_2 are weighed in the appropriate proportions. In the sintering process, the calcined powders are usually mixed in the desired shape.

M. Di Paolo Emilio, *Microelectronic Circuit Design for Energy Harvesting Systems*, DOI 10.1007/978-3-319-47587-5_5

The final baking process at high temperatures (approx. 1200 °C for 16 h for PZT) enables the ceramic to reach its optimum density. The polycrystalline piezoceramic materials must be polarized in an electric field in order to align the electric dipoles for improving the piezoelectric properties of the material. Moreover, the doping with small amounts of impurities can significantly improve the properties. Currently, many research efforts are aimed for optimizing the properties by selecting appropriate doping formulas. The most well-known difference is the distinction between the so-called hard and soft piezoceramic actuators. A material with a greater than 1 kV/mm field is called hard piezoelectric and a material with a pitch between 0.1 and 1 kV/mm is called soft piezoelectric. For electrical connection of the piezoelectric material, suitable electrode materials and related production processes have to be chosen. The most commonly process uses silver-palladium, which is sputtered or printed as a polymeric paste on the piezoelectric device. The piezoelectric effect of any chosen material is limited by its Curie or phase transition temperature. For the PZT, the Curie temperature varies between 250 °C and 400 °C and is depending on its composition.

5.3 Model

A piezoelectric material is a transducer which converts electrical energy into mechanical and vice versa. A schematic representation blocks can be displayed in Fig. 5.1. The electric lock is defined by two parameters: the intensity of the electric field E and the dielectric displacement D. That mechanical, instead, is represented by the mechanical solicitation T and the mechanical stress S.

The ratio of the gate parameters is mathematically described by the constitutive equations. The equations use the mechanical stress T and the electric field E as independent variables and is called the d-formulation. The same formula applies in the case of the material in energy harvesting area where T and D are the independent variables. In this case the expressions of interest are the following:

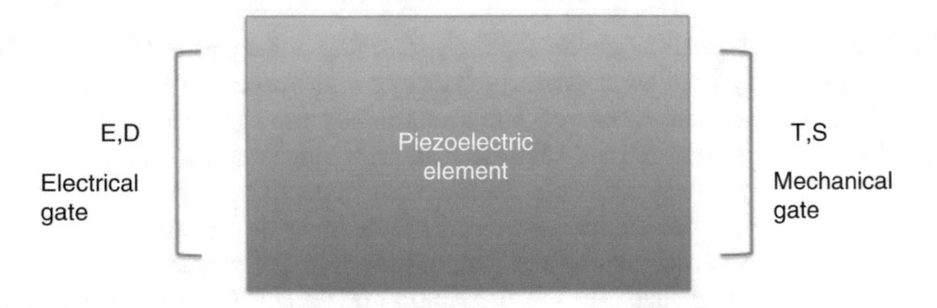

Fig. 5.1 General layout of a piezoelectric material

$$E = -gT + \frac{D}{\epsilon^T} \tag{5.1}$$

$$S = s^d T + gD \tag{5.2}$$

where s_d is the compliance (inverse of the Young modulus) measured with an electric charge on the electrodes kept constant (expressed in m/N); ϵ^T is the dielectric permittivity when a certain T (expressed in C/mV) is applied; g the coefficient of piezoelectricity expressed in Vm/N [16–20]. Piezoelectric materials are anisotropic materials for which constants are tensor and the electrical and mechanical variables of the gate are expressed with the vectors. It is demonstrated that the equations that describe the phenomenon are the following:

$$Q_p = C_p V_p - d_{33} F_p \tag{5.3}$$

$$z_p = d_{33} V_p - \frac{1}{k_p} F_p \tag{5.4}$$

where k_p is the stiffness of the material, F_p is the force of activation, z_p is the displacement in the respective direction, C the electric charge, d_{33} is parameter of the component D, Q_p is the charge, and V_p the applied or generated voltage. Whereas the electrical energy produced by the electromechanical response of the piezoelectric material with respect to the mechanical energy supplied to the material, the so-called coupling factor k is defined according to the following equation, in other words the ratio between the electrical power stored and the input mechanical:

$$k^2 = \frac{d^2}{\epsilon_0 \epsilon_T s} \tag{5.5}$$

The conversion of the vibration energy into electrical energy power is a crucial point for a successful design of energy harvesting devices. In general, a piezoelectric energy harvesting system is often modeled as a vibrating system driven and damped. This structure is constituted by a piezoelectric transducer coupled with the mechanical structure and connected to an energy storage system by an energy harvesting circuit. A mathematical level can be modeled with a set of N ordinary differential equations [21–30]. Whereas an energy collection system can be described as a two-port model with one degree of freedom (Fig. 5.2) due to the fact that an energy harvesting device is often tuned to a certain natural frequency. In this case the differential equations can be described by the following:

$$mu''(t) + bu'(t) + ku(t) + AV(t) = F(t) \tag{5.6}$$

$$-Au'(t) + C_p V'(t) = -I(t) \tag{5.7}$$

where u is the displacement of the mass m and k is the overall stiffness of the piezoelectric transducer and any other rigidity is connected. The effective

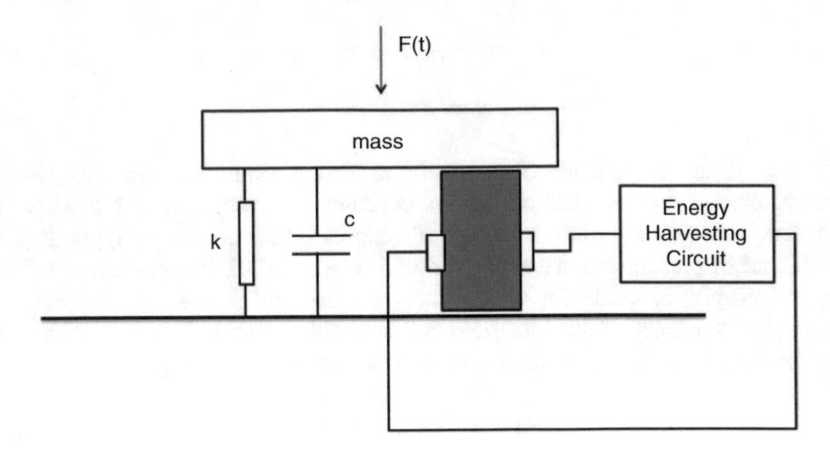

Fig. 5.2 General layout of a piezoelectric oscillator

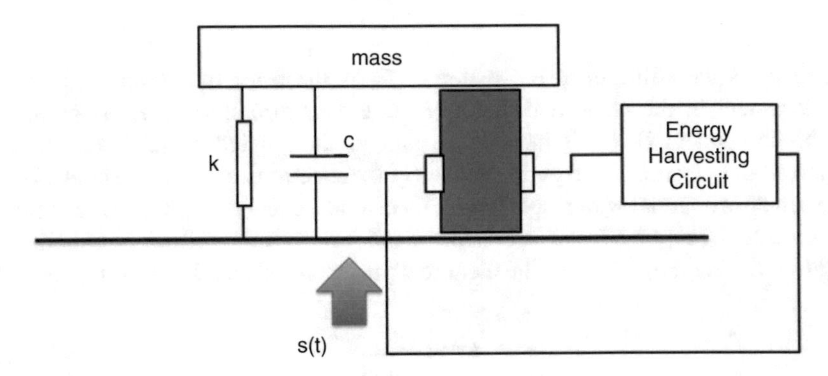

Fig. 5.3 General layout of a piezoelectric oscillator with a basic solicitation

piezoelectric coefficient A and capacities C_p depend on the geometry of the transducer and the load direction. Often, the energy collection system is applied over a vibrating mechanism, and in this configuration the modeling must be expressed whereas in Fig. 5.2 a basic solicitation. The system scheme is illustrated in Fig. 5.3. The equations which govern the system are the following:

$$mq''(t) + bq'(t) + kq(t) + AV(t) = -ms''(t) \tag{5.8}$$

$$-Aq'(t) + C_pV'(t) = -I(t) \tag{5.9}$$

In this case, $q(t) = u(t) - s(t)$ represents the difference displacement of the mass m and the excitation of the base. For a basic solicitation, the right side of the equation expressed by $-ms''(t)$ is a d'Alembert force induced by the acceleration of the ground. The equation of the system energy is obtained by multiplying the above equation for u for the expression of the tension $V(t)$. Piezoelectric devices are often modeled as current sources. The capacitance C_p of their inner electrodes is considered in parallel with the load resistor R. Assuming that the internal current generator is independent of the impedance of the external load, then the term V can be eliminated from the above equation. This assumption is equivalent to the statement that the connection is very weak or does not exist.

5.4 System Design

The voltage generated depends on the load R and the inductance L. It is assumed that the system is driven by a force of external excitation, which is sinusoidal with a frequency close to the natural frequency of the system with the piezoelectric element loaded. To improve the performance of energy harvesting systems with piezoelectric transducers can be required to apply more than one transducer on the structure. For practical reasons, the transducers cannot be connected to different collection circuits, but must be connected in series or in parallel (Fig. 5.4).

The piezoelectric transducers work as electrical generators and can be characterized by the equivalent circuit diagram of Fig. 5.4. C_p is the ability of the piezoelectric transducer and the electric current I_p is resulting from mechanical excitation of the piezoceramic [30–35]. For a piezo element of width b and length l, the electrical current can be written as the following in the plane stress hypothesis:

$$I_p(t) = \frac{d_{31}}{s_{11} + s12} bl \frac{d\epsilon}{dt} \tag{5.10}$$

Considering a sinusoidal excitation:

$$I_p(t) = Isin(\omega t) \tag{5.11}$$

The current and voltage values on the load can be expressed in conjugate complex form as follows:

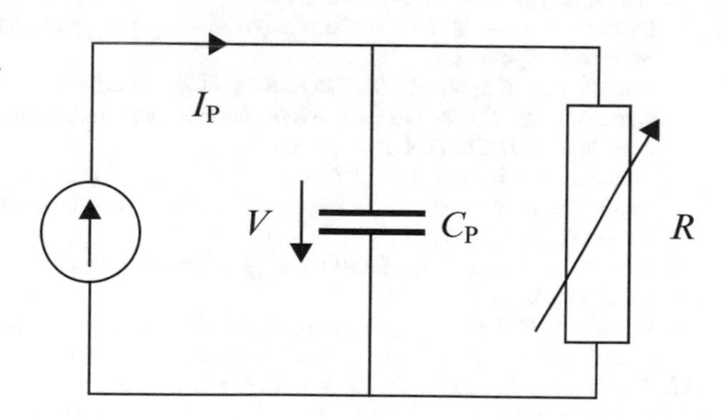

Fig. 5.4 Equivalent circuit for a piezoelectric transducer

$$I_r = \frac{1}{1 + j\omega RC_p} I \tag{5.12}$$

$$V_r = \frac{R}{1 + j\omega RC_p} I \tag{5.13}$$

By varying the resistance R, it is possible to find the value for which there is the maximum energy transfer. The power is expressed by the following equation:

$$P = \frac{I^2 R}{1 + (R\omega C_p)^2} \tag{5.14}$$

where we can define:

$$R_M = \frac{1}{\omega C_p} \tag{5.15}$$

To obtain the maximum power:

$$P_{\max} = \frac{I^2}{2\omega C_p} \tag{5.16}$$

References

1. Park, J., & Mackqy, S. (2003). *Practical Data Acquisition for instrumentation and system control*. Elsevier, Oxford.
2. Lacanette, K. (2003). *National temperature sensors handbook*. Annali di Matematica Pura ed Applicata. National Semiconductor.
3. National Instruments. (1996). *Data acquisition fundamentals, application note 007*.
4. National Instruments. (1996). *Signal conditioning fundamentals for PC-based data acquisition, Application Note 048*.
5. Taylor, J. (1986). *Computer-based data acquisition system*. Instrument Society of America.
6. Di Paolo Emilio, M. (2013). *Data acquisition system, from fundamentals to applied design*. New York: Springer.
7. Roundy, S., Wright, P., & Pister. K. (2002). Micro-electrostatic vibration-to- electricity converters. In *Proceedings of ASME international mechanical engineering congress and exposition (IMECE)* (Vol. 220, pp. 17–22).
8. Stordeur, M., & Stark, I. (1997). Low power thermoelectric generator: self-sufficient energy supply for micro systems. In *Proceedings of the 16th international conference on thermoelectrics* (pp. 575–577).
9. Shenck, N., & Paradiso, J. (2001). Energy scavenging with shoe-mounted piezoelectrics. *Micro IEEE, 21*(3), 30–42.
10. Roundy, S. (2003). *Energy scavenging for wireless sensor nodes with a focus on vibration to electricity conversion*. Ph.D Thesis, University of California.
11. Tsutsumino, T., Suzuki, Y., Kasagi, N., Kashiwagi, K., & Morizawa, Y. (2006). *Micro seismic electret generator for energy harvesting*. Technical Digest PowerMEMS (pp. 133–136). Berkeley, USA.

12. Sterken, T., Altena, G., Fiorini, P., & Puers, R. (2007). *Characterisation of an electrostatic vibration harvester, EDA Publishing Association.*
13. Sterken, T., Baert, K., Puers, R., & Borghs, S. (2002). Power extraction from ambient vibration. In *Proceedings of the SeSens (Workshop on Semiconductor Sensors, Veldhoven, Netherlands)* (pp. 680–683).
14. Szarka, G., Stark, B., & Burrow, S. (2012). Review of power management for energy harvesting systems. *IEEE Transactions on Power Electronics, 27*(2), 803–815. ISSN: 0885-8993.
15. Cammarano, A., Burrow, S. G., Barton, D. A. W., Carrella, A., & Clare, L. R. (2010). Tuning a resonant energy harvester using a generalized electrical load. *Smart Materials and Structures, 19*, 055003.
16. Guyomar, D., Badel, A., Lefeuvre, E., & Richard, C. (2005). Toward energy harvesting using active materials and conversion improvement by nonlinear processing. *IEEE Transactions on Ultrasonics, Ferroelectrics, and Frequency Control, 52*, 584–595.
17. Mitcheson, P. D., Stoianov, I., & Yeatman, E. M. (2012). Power-extraction circuits for piezoelectric energy harvesters in miniature and low-power applications. *IEEE Transactions on Power Electronics, 27*, 4514–4529.
18. Szarka, G. D., Burrow, S. G., & Stark, B.H. (2012). Ultra-low power, fully-autonomous boost rectifier for electro-magnetic energy harvesters. *IEEE Transactions on Power Electronics, 28*(7), 3353–3362. doi:10.1109/TPEL.2012.2219594.
19. Maurath, D., Becker, P. F., Spreeman, D., & Manoli, Y. (2012). Efficient energy harvesting with electromagnetic energy transducers using active low-voltage. *IEEE Journal of Solid-State Circuits, 47*(6), 1369–1380.
20. Beeby, S. P., Tudor, M. J., & White, N. M. (2006). Energy harvesting vibration sources for microsystems applications. *Measurement Science and Technology, 17*, R175–R195.
21. Khaligh, A., Zeng, P., & Zheng, C. (2010). Kinetic energy harvesting using piezoelectric and electromagnetic technologies—state of the art. *IEEE Transactions on Industrial Electronics, 57*(3), 850–860.
22. Paulo, J., & Gaspar, P. D. (2010). Review and future trend of energy harvesting methods for portable medical devices. In *Proceedings of the World Congress on Engineering* (Vol. 2)
23. Zhu, D., Tudor, M. J., Beeby, S. P. (2010). Strategies for increasing the operating frequency range of vibration energy harvesters: A review. *Measurement Science and Technology, 21*, 022001-1–022001-29.
24. Cepnik, C., Lausecker, R., & Wallrabe, U. (2013). Review on electrodynamic energy harvesters—a classification approach. *Micromachines, 4*(2), 168–196. http://www.mdpi.com/2072-666X/4/2/168. Accessed 20 Jan 2015.
25. Ulaby, F. T., Michielssen, E., & Ravaioli, U. (2010). *Fundamentals of Applied Electromagnetics* (6th ed.). Prentice Hall, USA.
26. Roundy, S., Wright, P. K., & Rabaey, J. M. (2003). A study of low level vibrations as a power source for wireless sensor nodes. *Computer Communications, 26*(11), 1131–1144.
27. Sazonov, E., Li, H., Curry, D., & Pillay, P. (2009). Self-powered sensors for monitoring of highway bridges. *IEEE Sensors Journal, 9*, 1422–1429.
28. Toh, T. T., Mitcheson, P. D., Holmes, A. S., & Yeatman, E. M. (2008). A continuously rotating energy harvester with maximum power point tracking. *Journal of Micromechanics and Microengineering, 18*, 104008-1-7.
29. Howey, D. A., Bansal, A., & Holmes, A. S. (2011). Design and performance of a centimetre-scale shrouded wind turbine for energy harvesting. *Smart Materials and Structures, 20*, 085021.
30. Razavi, B. (2002). *Design of analog CMOS integrated circuits*. McGraw-Hill
31. Razavi, B. (2008). *Fundamentals of microelectronics*. New York: Wiley.
32. Sedra, A. S., & Smith, K. C. (2013). *Microelectronic circuits*. Oxford: Oxford University.
33. Razavi, B. (2002). *Design of integrated circuits for optical communications*. McGraw-Hill.
34. Hurst, P. J. (2001). *Analysis and design of analog integrated circuits*. New York: Wiley.
35. Spies, P. (2015). *Handbook of energy harvesting power supplies and applications*. CRC Press Book, France.

Chapter 6
Thermoelectric Transducers

6.1 Introduction

The discovery of the thermoelectric has been observed by Thomas J. Seebeck in 1821 after the deviation of a compass near two metallic conductors connected each other at different temperatures. The degree of deflection was proportional to the temperature difference. The reason was to study the difference of the electric field that was created due to the temperature difference between the conductors. The effect observed by the Seebeck is reversible as described by C. A. Jean Peltier in 1834: if we supply electrical energy in two connected conductors, a temperature gradient occurs in the contact point; the thermal energy is transported from a point of connection to the other, leading a cooling effect [1–15].

6.2 Seebeck and Peltier Effect

The Peltier effect is based on the production or absorption of heat at a junction between two different conductors when an electrical charge flows through it. The dQ/dt rate of heat produced or absorbed at a junction between the conductors A and B is the following:

$$\frac{dQ}{dt} = (\alpha_a - \alpha_b)I \tag{6.1}$$

where I is the electrical current and α are the Peltier coefficients of the conductors. The Seebeck effect is the production of electromotive force between junctions of two different conductors. Two nodes connected back to back work with two different temperatures, T_H and T_C and a tension between their free contacts:

$$V = -S(T_h - T_c) \tag{6.2}$$

M. Di Paolo Emilio, *Microelectronic Circuit Design for Energy Harvesting Systems*, DOI 10.1007/978-3-319-47587-5_6

S is the Seebeck coefficient. The Thomson effect is the production or absorption of heat along a conductor with a temperature gradient ΔT when the electric charge flows through it. The heat dQ/dt produced or absorbed along a conductor segment is the following:

$$\frac{dQ}{dt} = -KJ\Delta T \quad (6.3)$$

where J is the current density, K is the Thomson coefficient. The coefficients are governed by the following relations:

$$\alpha = TS \quad (6.4)$$

with

$$\alpha = \alpha_a - \alpha_b \quad (6.5)$$

and

$$K = T\frac{dS}{dT} \quad (6.6)$$

Although these main thermoelectric effects have been known for a long time, it is difficult to find explicit expressions in the literature for their three coefficients in terms of the most fundamental physical quantity. The electrons in the wires occupy energy levels in pairs of opposite spin. The lower levels are fully occupied and higher levels are empty and the population of the level is determined by the Fermi-Dirac statistics [16–25].

To move in the conductor an electron is occupying a certain level, it must be dispersed to a vacuum level. For this reason, the low-energy electrons do not contribute to the electric current because their neighboring levels are occupied. The Fermi level is the energy at which the probability of occupation electron is 0.5. Only electrons with energies near this level contribute to the current. The average kinetic energy of the particles is calculated by adding all the velocity squared over all directions in space in a solid angle of 4π determined by an angle θ, from 0 to π, and azimuth of 2π. Each speed is weighed by a probability distribution $f_0(x, v, \theta)$:

$$< v^2 > \int f_0(x, v, \theta)|v|^2 d^3v \quad (6.7)$$

where $d^3v = 2\pi v 2dv \sin(\theta)d\theta$. The Maxwell–Boltzmann distribution is the following:

$$f_0(v) = \frac{m}{2\pi kT}\frac{3}{2}e^{-mv^2/2KT} \quad (6.8)$$

where the exponential factor of last equation is determined by the following condition:

$$\int_0^\infty f_0 4\pi v^2 dv = 1 \tag{6.9}$$

and it is applied to the calculation of $< v^2 >$. The average kinetic energy is the following:

$$< v^2 >= 4\pi \int f_0(v) v^4 dv = 3KT/m \tag{6.10}$$

$$\frac{1}{2} m < v^2 >= \frac{3}{2} kT \tag{6.11}$$

6.3 Potential

As example we can consider two pieces of N-type semiconductor (L and R) with n_L more than n_R, where n_L and n_R are equal to the corresponding donors density. Therefore, these densities are independent from temperature. If the two pieces are brought in contact to form a junction, electrons will start to spread. The diffusion creates a depletion region with electric field and the potential difference V_c between the two pieces which stops the further diffusion of electrons. V_c is the potential for contact expressible by the following formula:

$$V_c = \frac{kT}{e} ln(n_L/n_R) \tag{6.12}$$

where e is the electron charge and k the Boltzmann constant.

6.4 Charges in a Semiconductor with a Temperature Gradient

The electrical current through a semiconductor with a temperature gradient can be calculated by applying the transport equation of the Boltzmann. This equation produces expressions for the currents that are similar to the linear equations of Onsager, but with the advantage that there are no unknown linear coefficients. If the density of the charge carriers is not dependent on position, then the electric current will be:

$$J_q = \sigma \left(E - \frac{\frac{k}{2e} dT}{dx} \right) \tag{6.13}$$

where σ is the electrical conductivity and E the electric field. By integrating above equation and to equal J_q to zero:

$$V = \frac{k}{2e}\Delta T \tag{6.14}$$

Namely, a voltage of 43 $\mu V/°C$ independent of the charge density and developed between the hot and cold spots.

6.5 Thermoelectric Effect

The original Carnot cycle involves the transformation of heat between a hot bath T_H and a T_C cold bath from a gas in a cylinder with a movable piston. The heat flows in part from a hot bath to a cold bath, and in part is converted into mechanical work in a closed cycle to four reversible stages. Carnot knew nothing of the chemical potential, even on entropy. Nevertheless, he studied the cycle correctly without these terms. At each stage of the macroscopic mechanical cycle, the work of the movable piston is transformed to the thermal motion of the gas particles by elastic collisions between the particles and the piston. Similarly, the electric work of a power source is converted into a microscopic thermal heat junction by the acceleration of the charge carriers in the electric field [26–30].

The electrons that pass through a junction will be slowed down or accelerated by a contact potential difference V_c. Thus, they absorb or provide a quantity of heat at the junction eV_C, where V_c is calculated above with Peltier coefficients given by the following equations:

$$\alpha_a = kTln(n_L) \tag{6.15}$$

$$\alpha_b = kTln(n_R) \tag{6.16}$$

6.6 Thomson Effect

The Thomson effect refers to the generation of heat by resulting from the passage of a current along a portion of a single conductor on which is applied a difference of temperature ΔT. Because of the temperature difference, absorbed heat per unit of time is given by the following equation:

$$\frac{dQ}{dt} = \beta I \Delta T \tag{6.17}$$

where β is the coefficient of Thomson. The origin of the effect is substantially the same as for the Peltier effect. Here the temperature gradient along the conductor is responsible for differences in potential energy of the charge carriers. The Thomson effect is not of primary importance in thermoelectric devices, but it should not be overlooked by detailed calculations.

6.7 Thermoelectric Generator

A thermocouple usually consists—as the name suggests—of two different metals or alloys. When the two junctions are at different temperatures, a low voltage—of the order of tens of mV/K—is generated. This is called the Seebeck effect described in this chapter and that can be used in temperature measurement and control.

In thermocouples, for the generation of energy, are partly replaced by semiconductor granules, contacted with a highly conductive metal strips. With such materials, an order of magnitude greater are achieved and can also be used in thin film devices for the measurement of temperature. Therefore the thermocouple layers of micrometric thickness are deposited by a variety of techniques on a thin substrate of electrically insulating support film. A thermoelectric converter obeys the laws of thermodynamics; efficiency is defined as the ratio between the electrical power supplied to the load and the absorbed heat to the hot junction (Figs. 6.1 and 6.2).

Expressions for the important parameters in the thermal power generation can easily be deduced by considering the generator composed of a single thermocouple and n- and p-type semiconductors as shown in Fig. 6.2. The thermocouples are

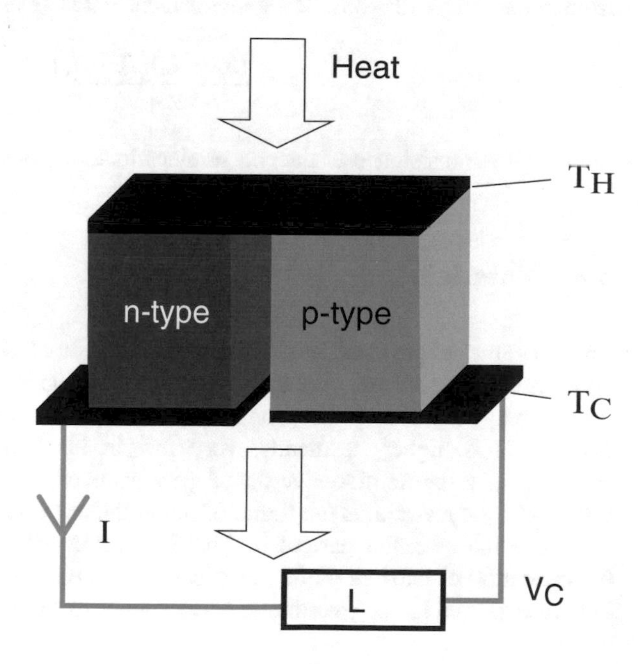

Fig. 6.1 General layout of a thermocouple

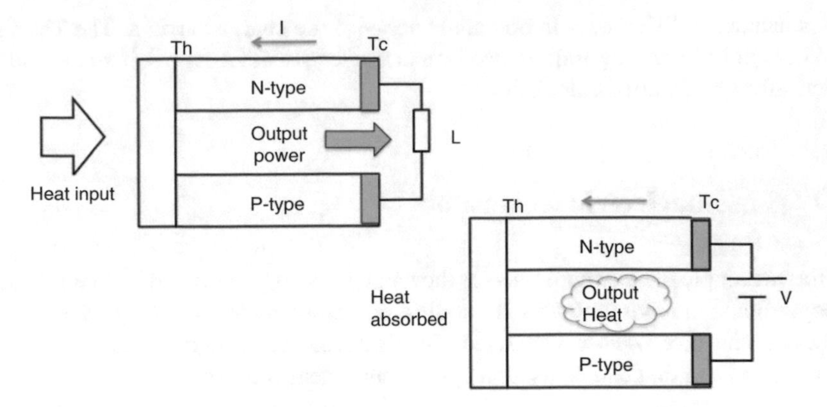

Fig. 6.2 Thermoelectric generator

constructed by two branches, one of n-type, a p-type material with the length L_n and L_p, and constant cross sections A_n and A_p. The two branches are connected to metal conductors of negligible electrical resistance. The heat is transferred only by conduction along the branches of the thermocouple. The thermocouple is used as a current generator by means of Seebeck or Peltier effect. The efficiency of a generator is defined as the ratio between the energy provided to the load and the heat absorbed by the junction. The load energy is related to the current and the resistance itself; the transported heat instead is only linked by thermal conduction and then the conductivity parameters of the two materials, with another dependence from Peltier effect [31–35]. The generated current is given by:

$$I = \frac{(S_p - S_n)(T_h - T_c)}{R + R_l} \tag{6.18}$$

where R is the resistance of the two semiconductors blocks of n and p.

6.8 Materials

Three parameters are used for classification of thermoelectric materials: σ, electrical conductivity; λ, thermal conductivity, and the Seebeck coefficient. The electrical conductivity is given by the product of the concentration and mobility of charge carriers. It is higher in metals, very low in insulators, with an intermediate position taken by semiconductors. As measure of the potential usefulness of a thermoelectric material is the figure of merit, the three parameters above mentioned constitute the essential part of it. The Seebeck coefficient drops with increasing concentration of carriers while electrical conductivity increases; consequently, the electrical power factor (parameter proportional to the conductivity multiplied by the Seebeck coefficient squared) has a maximum that is typically located around a carrier concentration of about $10^{19}/\text{cm}^3$. There are two components of the thermal conductivity: vibration pattern and the electronic part. The latter also increases with concentration of carriers and typically accounts for about one third of the thermal conductivity. The maximum amount of energy falls into the region of semiconductors. Therefore, the semiconductors are the materials of choice for the further development of thermoelectric devices. Thermoelectric devices were further classified with respect to the temperature ranges to which they can be usefully employed. A positive direction of development has been the reduction of the thermal conductivity, another search for the so-called electronic crystal glass phonon, in which it is assumed that the crystal structures with weakly bound atoms or molecules inside an atomic cage should conduct heat like a glass, but conduct electricity like a crystal. Over the past decade, materials scientists were optimistic in their belief that the low-dimensional structures such as quantum wells (materials that are so thin as to be essentially two-dimensional 2D), quantum wires (very small section and considered to be in one dimension 1D and referred to as nano-wires), quantum dots that are confined in all directions, and superlattices (a multi-layer structure of quantum wells) will provide a path for the achievement of a significant improvement of figures of merit. There are also ongoing attempts to improve the competitiveness of thermoelectric materials in directions different from those of the figure of merit, such as reduction of costs, to the development of more environmentally friendly materials.

6.9 Figure of Merit

Typically, an n-type and a p-type thermoelectric material are arranged thermally in parallel and electrically in series as shown in Fig. 6.2. An effective thermoelectric figure of merit for the two materials used in a module can be defined as follows:

$$ZT = \frac{(S_p - S_n)^2}{(\sqrt{\lambda_p \rho_p} + \sqrt{\lambda_n \rho_n})^2} \tag{6.19}$$

where λ is the thermal conductivity and ρ the resistivity of the material. By means of demonstrations, it can be shown that the maximum power is generated when the external load resistance corresponds to the internal electrical resistance of the pair. At this operating point, the power produced is given by the following expression:

$$W = \frac{((S_p - S_n)(T_h - T_c))^2}{4R} \tag{6.20}$$

At this point of maximum power, the power efficiency can be approximated as follows:

$$\eta = \frac{Z\Delta T}{4 + ZT_h + ZT_m} = \frac{\Delta T}{T_h} \frac{1}{2 + \frac{4}{ZT_h} - \frac{\Delta T}{2T_h}} \quad (6.21)$$

In this case ZT is calculated as function of the average temperature.

References

1. Park, J., & Mackqy, S. (2003). *Practical Data Acquisition for instrumentation and system control*. Elsevier, Oxford.
2. Lacanette, K. (2003). *National temperature sensors handbook*. Annali di Matematica Pura ed Applicata. National Semiconductor.
3. National Instruments. (1996). Data acquisition fundamentals, Application note 007.
4. National Instruments. (1996). *Signal conditioning fundamentals for PC-based data acquisition*, Application Note 048
5. Taylor, J. (1986). *Computer-based data acquisition system*. Instrument Society of America.
6. Di Paolo Emilio, M. (2013). *Data acquisition system, from fundamentals to applied design*. New York: Springer.
7. Roundy, S., Wright, P., & Pister. K. (2002). Micro-electrostatic vibration-to- electricity converters. In *Proceedings of ASME international mechanical engineering congress and exposition (IMECE)* (Vol. 220, pp. 17–22).
8. Stordeur, M., & Stark, I. (1997). Low power thermoelectric generator: self-sufficient energy supply for micro systems. In *Proceedings of the 16th international conference on thermoelectrics* (pp. 575–577).
9. Shenck, N., & Paradiso, J. (2001). Energy scavenging with shoe-mounted piezoelectrics. *Micro IEEE, 21*(3), 30–42.
10. Roundy, S. (2003). *Energy scavenging for wireless sensor nodes with a focus on vibration to electricity conversion*. Ph.D Thesis, University of California.
11. Tsutsumino, T., Suzuki, Y., Kasagi, N., Kashiwagi, K., & Morizawa, Y. (2006). *Micro seismic electret generator for energy harvesting*. Technical Digest PowerMEMS (pp. 133–136). Berkeley, USA.
12. Sterken, T., Altena, G., Fiorini, P., & Puers, R. (2007). *Characterisation of an electrostatic vibration harvester*, EDA Publishing Association.
13. Sterken, T., Baert, K., Puers, R., & Borghs, S. (2002). Power extraction from ambient vibration. In *Proceedings of the SeSens (Workshop on Semiconductor Sensors, Veldhoven, Netherlands)* (pp. 680–683).
14. Szarka, G., Stark, B., & Burrow, S. (2012). Review of power management for energy harvesting systems. *IEEE Transactions on Power Electronics, 27*(2), 803–815. ISSN: 0885-8993.
15. Cammarano, A., Burrow, S. G., Barton, D. A. W., Carrella, A., & Clare, L. R. (2010). Tuning a resonant energy harvester using a generalized electrical load. *Smart Materials and Structures, 19*, 055003.
16. Guyomar, D., Badel, A., Lefeuvre, E., & Richard, C. (2005). Toward energy harvesting using active materials and conversion improvement by nonlinear processing. *IEEE Transactions on Ultrasonics, Ferroelectrics, and Frequency Control, 52*, 584–595.
17. Mitcheson, P. D., Stoianov, I., & Yeatman, E. M. (2012). Power-extraction circuits for piezoelectric energy harvesters in miniature and low-power applications. *IEEE Transactions on Power Electronics, 27*, 4514–4529.

18. Szarka, G. D., Burrow, S. G., & Stark, B.H. (2012). Ultra-low power, fully-autonomous boost rectifier for electro-magnetic energy harvesters. *IEEE Transactions on Power Electronics, 28*(7), 3353–3362. doi:10.1109/TPEL.2012.2219594.
19. Maurath, D., Becker, P. F., Spreeman, D., & Manoli, Y. (2012). Efficient energy harvesting with electromagnetic energy transducers using active low-voltage. *IEEE Journal of Solid-State Circuits, 47*(6), 1369–1380.
20. Beeby, S. P., Tudor, M. J., & White, N. M. (2006). Energy harvesting vibration sources for microsystems applications. *Measurement Science and Technology, 17*, R175–R195.
21. Khaligh, A., Zeng, P., & Zheng, C. (2010). Kinetic energy harvesting using piezoelectric and electromagnetic technologies—state of the art. *IEEE Transactions on Industrial Electronics, 57*(3), 850–860.
22. Paulo, J., & Gaspar, P. D. (2010). Review and future trend of energy harvesting methods for portable medical devices. In *Proceedings of the World Congress on Engineering* (Vol. 2)
23. Zhu, D., Tudor, M. J., Beeby, S. P. (2010). Strategies for increasing the operating frequency range of vibration energy harvesters: A review. *Measurement Science and Technology, 21*, 022001-1–022001-29.
24. Cepnik, C., Lausecker, R., & Wallrabe, U. (2013). Review on electrodynamic energy harvesters—a classification approach. *Micromachines, 4*(2), 168–196. http://www.mdpi.com/2072-666X/4/2/168. Accessed 20 Jan 2015.
25. Ulaby, F. T., Michielssen, E., & Ravaioli, U. (2010). *Fundamentals of Applied Electromagnetics* (6th ed.). Prentice Hall, USA.
26. Roundy, S., Wright, P. K., & Rabaey, J. M. (2003). A study of low level vibrations as a power source for wireless sensor nodes. *Computer Communications, 26*(11), 1131–1144.
27. Sazonov, E., Li, H., Curry, D., & Pillay, P. (2009). Self-powered sensors for monitoring of highway bridges. *IEEE Sensors Journal, 9*, 1422–1429.
28. Toh, T. T., Mitcheson, P. D., Holmes, A. S., & Yeatman, E. M. (2008). A continuously rotating energy harvester with maximum power point tracking. *Journal of Micromechanics and Microengineering, 18*, 104008-1-7.
29. Howey, D. A., Bansal, A., & Holmes, A. S. (2011). Design and performance of a centimetre-scale shrouded wind turbine for energy harvesting. *Smart Materials and Structures, 20*, 085021.
30. Razavi, B. (2002). *Design of analog CMOS integrated circuits*. McGraw-Hill
31. Razavi, B. (2008). *Fundamentals of microelectronics*. New York: Wiley.
32. Sedra, A. S., & Smith, K. C. (2013). *Microelectronic circuits*. Oxford: Oxford University.
33. Razavi, B. (2002). *Design of integrated circuits for optical communications*. McGraw-Hill.
34. Hurst, P. J. (2001). *Analysis and design of analog integrated circuits*. New York: Wiley.
35. Spies, P. (2015). *Handbook of energy harvesting power supplies and applications*. CRC Press Book, France.

Chapter 7
Electrostatic Transducers

7.1 Introduction

Micro-generators have applications in wireless sensor networks for security systems, military applications, the monitoring of structural conditions, and personal health systems; they require work to be done on a transducer so that electricity can be generated. This requires a relative movement between the two ends of the transducer (referred to the rotor and stator in the conventional generators) and it is established by fixing the stator and allowing the movement of the rotor by connecting it directly to the source, or through the use of an inertial mass, as shown in Fig. 7.1. The inertial method is preferred because a single point of attachment is necessary to collect energy from the source [1–15].

The system mainly includes an inertial mass of test, a spring and two shock absorbers. One of these, D_e, is the mechanism of conversion between electric energy and kinetic energy, and the other represents the parasitic mechanical damping. The amount of energy that can be generated under specific operating conditions in terms of amplitude and vibration is strongly dependent on the electrical damping. In addition, the damping force must be tuned to enable the generator to continue to work in maximum efficiency. One of the major design decisions is that the type of transduction mechanism should be used. The electrostatic generators can make use of the variable capacitor structures or piezoelectric materials. The choice of the type of transducer for a particular application of generator is made on the basis of the following two main criteria:

- Compatibility MEMS: The potential for widespread use of powered devices means that they must be cheap to produce market. They are always embedded with some electronic sensing and data transmission. Consequently, a very useful characteristic of a micro-generator is its ability to be produced in large quantities at low cost and that can be easily integrated with the electronics.

M. Di Paolo Emilio, *Microelectronic Circuit Design for Energy Harvesting Systems*, DOI 10.1007/978-3-319-47587-5_7

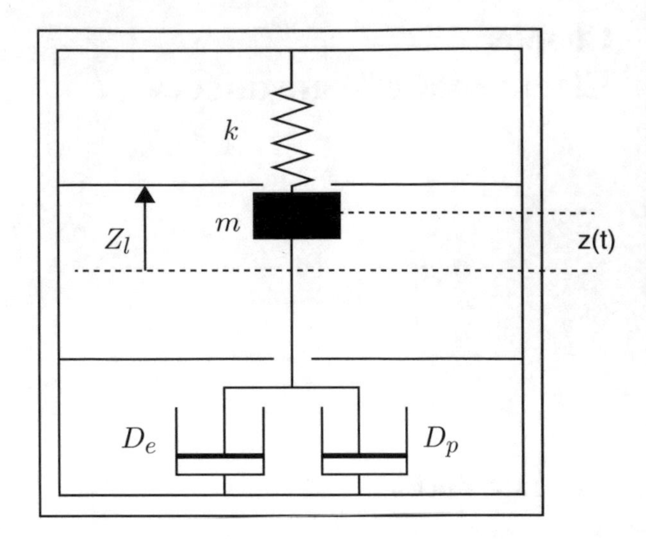

Fig. 7.1 Inertial generator

- Controllability: Large-scale energy harvesting devices, such as photovoltaic panels and wind turbines maximum, use of technical system about power point tracking in order to collect as much energy as possible. An important aspect of micro-generators is that, in order to maintain a high degree of efficiency, it should be possible to apply techniques of monitoring (sensing).

7.2 Physical Phenomena

The electrostatic energy conversion mechanism of a transducer consists in the physical coupling between electricity and mechanical means of an electrostatic force. The electrostatic force is induced between opposite charges stored on two opposite electrodes. The amount of charge Q accumulates on the electrodes is a function of the potential difference V between the electrodes and the capacity C according to the relationship $Q = CV$. The stored energy of the capacitor is controlled by the following formula:

$$E = \frac{1}{2}CV^2 \tag{7.1}$$

The physical principle of energy conversion cycle depends on the conditioning circuit common to all energy harvesting generators and how the variable capacitor is connected to the corresponding electrical circuits. In general we can distinguish two types of connection: switching and continuous systems.

7.3 Switching System

The switched connection between the transducer and the circuit involves a reconfiguration of the system, through the operation of switch in different phases of the generation cycle. The switched transducers can be further divided into two main types: constant charge and constant voltage. In the constant charge if a variable capacitor is pre-loaded to maximum capacity and then disconnected from any external circuit before the capacitor geometry is modified by the motion, then the extra energy will be stored in the electric field between the electrodes as a work against the force electrostatic [16–25]. This energy can then be used to power a circuit. The most common way in which this approach is implemented is shown in Fig. 7.2.

The device is pre-charged to a low voltage in the first part of the cycle as shown in the diagram QV of Fig. 7.2. The plates are arranged so that they can be separated, thereby increasing the distance between them. The movement produces a constant force between the two electrodes. The area outlined by QV diagram represents the electricity generated. In the constant voltage mode, the capacitor is connected to a constant voltage (possibly supplied by a battery), a reduction of capacity between the electrodes caused by the relative motion of the plates would lead to the charge to be removed from the condenser and pushed back into the voltage source, thereby increasing the energy stored. If the plates are actuated in a sliding motion, as indicated in Fig. 7.3, the force between the plates in the direction of relative movement remains constant. The electronic circuitry to achieve a switching of the transducer system can be quite different depending on the type of energy conversion cycle. In Figs. 7.4 and 7.5 is shown an implementation of the circuit for each mode of operation.

The switches are realized by MOSFET devices, which are operated by some electronic control. The capacitor CV denotes the variable capacitor, while CR denotes a storage capacitor (CRL: low voltage, CRH: high voltage). In the constant voltage mode, the circuit must perform three different tasks: charge the variable

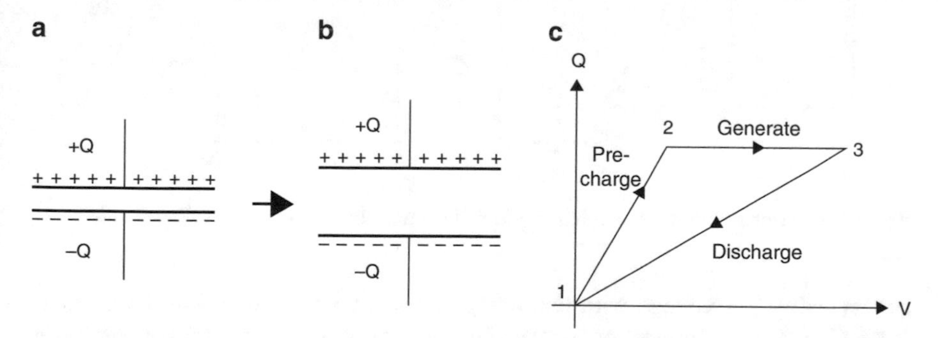

Fig. 7.2 Operation of an electrostatic transducer in constant charge mode. In (**a**) and (**b**) the operational conditions and (**c**) the diagraph QV

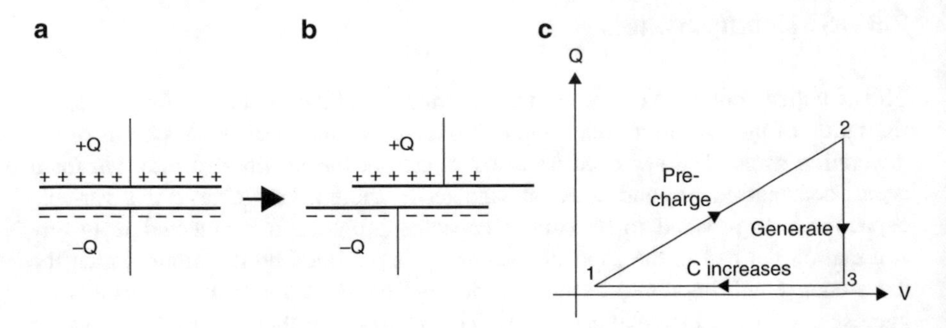

Fig. 7.3 Operation of an electrostatic transducer in constant voltage mode. In (**a**) and (**b**) the operational conditions and (**c**) the diagraph QV

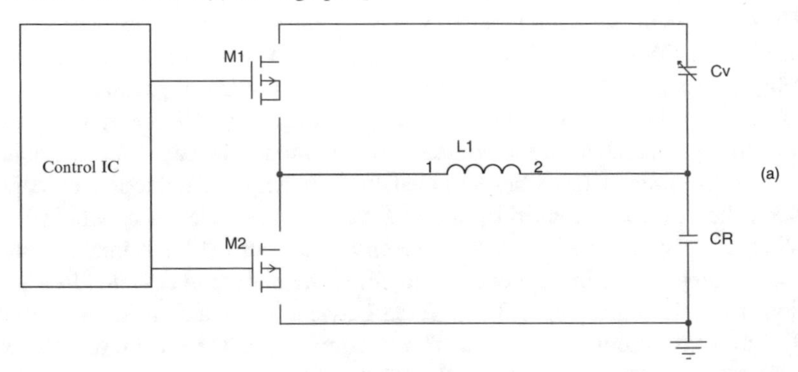

Fig. 7.4 Electronic circuit for a switched electrostatic transducer in constant charge mode

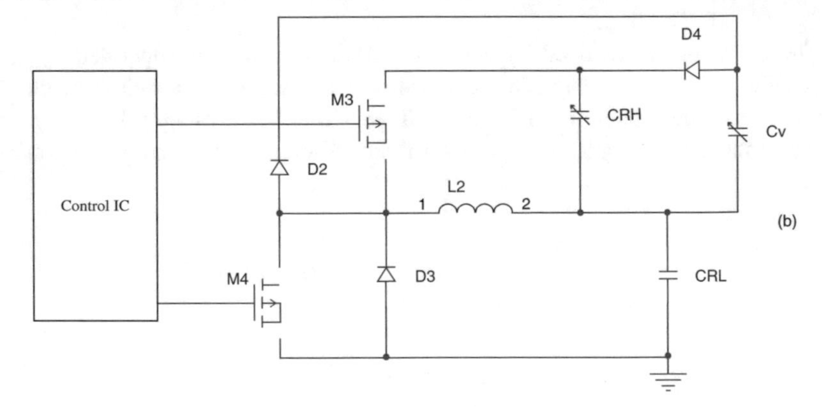

Fig. 7.5 Electronic circuit for a switched electrostatic transducer in constant voltage mode

capacitor to V_{high} voltage with the voltage constant and by reducing the capacity from C_{max} to C_{min}; moreover by transferring energy from the high-voltage to the low-voltage store. In the other mode the associated circuit must implement the following functions: charge the capacitor to a voltage level of V_{low} by maintaining constant the charge in the capacitor while the capacity is reduced from C_{max} to C_{min}, and the charge transfer from the variable capacitor to a store. The constant charging phase is achieved by simply unplugging the variable capacitor by using electronic or mechanical switches.

7.4 Continuous Systems

A third mode of operation is when the variable capacitor is constantly connected to the load circuit, and this load circuit provides a bias voltage to the capacitor. A simple example of this is a voltage source, a resistor, and a variable capacitor connected in series. A capacitance change will always result in a charge transfer between the electrodes through the load resistor by causing a work to be performed in the load through a transfer of energy. A constant charge generator is equivalent to a continuous generator with an infinite impedance load, while the constant voltage generator corresponds to a continuous generator short-circuited. The use of controlled switches complicates the implementation of the generator and the circuitry necessary to control by consuming a minimal amount of power generated, and therefore in some cases a continuous system is preferred over the other two systems. The design can be based on a variable capacitor with a constant source or time dependent [26–30]. A well-known example of the use of bias built-in sources is given by piezoelectric generators: the capacitance between the electrodes of the generator is therefore practically constant, but the bias voltage variations change as a function of the displacement (Fig. 7.6).

If the piezoelectric materials are not available or are not applicable to the target application, electrets materials can be used. In the silicon-based technologies it is appropriate to use an SiO2/Si3N4 bilayer, similar to those used for the flash memory cells. If the use of these layers is not compatible with the design of the capacitor, then polymers such as Teflon or parylene are applied. In a continuous system, the

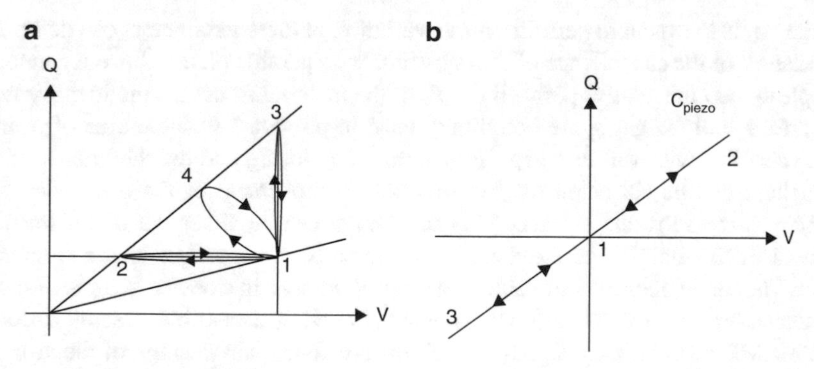

Fig. 7.6 Operation mode of a capacitor in continuous mode (**a**) or with a piezoelectric (**b**)

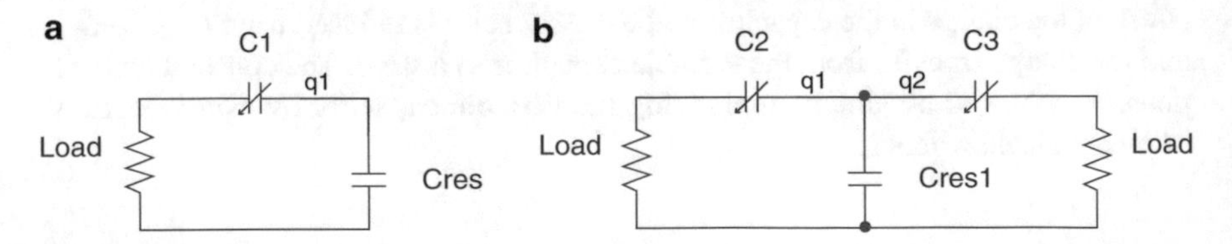

Fig. 7.7 Electrical scheme for the operation of the electrostatic transducer in continuous mode

variable capacitor is constantly connected to the circuit which includes the load. A capacitance change will always result in a transfer of charge through the load resistance. An advantage of a continuous system is that the transducer system can be implemented without the use of switches. The use of switches requires some extra circuitry to control them and valuable energy is consumed by the control circuit. There are two basic schemes to achieve a continuous system. In a pattern, a single variable capacitor is used in series with a voltage generator (provides the bias voltage) and a load resistor (Fig. 7.7a). In this case, the charge flows through the voltage source. An alternative method (Fig. 7.7b) implements two additional capacitors with its capacity that varies in the opposite way. One of the advantages of the latter method is that transduction is quite insensitive to parasitic capacitances.

7.5 Design

The simple design is a parallel-plate capacitor. The ability of a parallel-plate capacitor is determined by the area A and the distance g of the plates and by the dielectric E_r of the material between the plates:

$$C = \frac{A}{g}\epsilon_0 E_r \tag{7.2}$$

where ϵ_0 is the vacuum permittivity. A variation of these parameters can determine a variation of the capacitance of the capacitor with parallel plates. Since it cannot be simple to vary the relative permittivity E_r of the material from the kinetic movement, the area A and the gap g are commonly used to provide a variable area of overlap or a capacitor with variable gap. In general, gap-closing and overlap-area variable capacitors can be classified by the direction of movement of the electrodes with respect to the substrate surface. This movement can be either flat or off level. In both cases the mobile and fixed electrodes must be electrically isolated from each other. The manufacture of movable and fixed electrodes in a device layer is based on silicon technology on insulator (SOI), which is widely accessible in semiconductor and MEMS technologies [2, 31–35]. There are four main designs of electrostatic transducers to implement miniature generators:

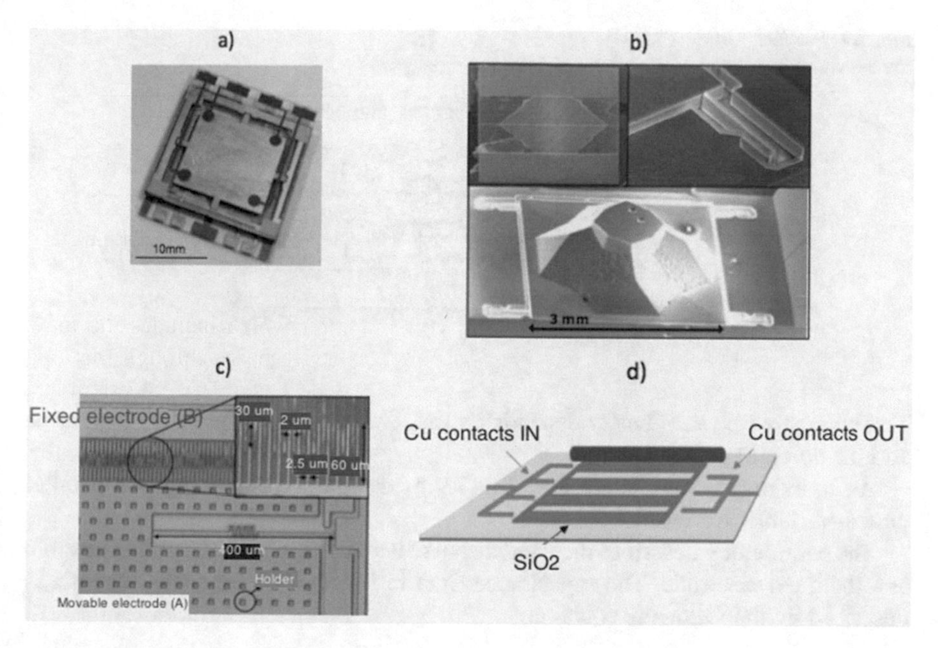

Fig. 7.8 Four types of electrostatic transducers: (**a**) parallel plate—gap closing, (**b**) parallel plate—variable overlap, (**c**) comb drive, (**d**) rolling rod

- Parallel-plate gap closing;
- Parallel-plate variable overlap;
- Comb drive;
- Rolling rod.

The device shown in Fig. 7.8a is a drawing to parallel plates, where they form a capacitance between two parallel electrodes stacked one above the other. The device is used in a constant charge mode and generates high-voltage spikes when it is primed with a low voltage. The device of Fig. 7.8b is a drawing to parallel plates and is generally combined with electrets and works in continuous mode. The third type of transducer is still composed of multiple capacitors in parallel, but this time the electrodes are on the same slice, forming one comb unit. The mass of the generator can move in the plane when the movement occurs, by changing the overlap of the electrodes. They may be operated continuously or in a switched mode. It can be shown that the maximum output power possible from these systems is proportional to the value of the test mass. If these systems are to be made in a cost-effective way should be achievable by using batch production with MEMS silicon standard. The device is shown in Fig. 7.8d and uses a proof mass that is not fixed to the frame, and an outer mass can be used to not incur the problems associated with the alignment and fastening of an external mass to the MEMS device. This structure is composed of a metal rod on dielectric layers, which serves as a movable electrode, and a series of strip electrodes beneath the dielectric as static electrodes.

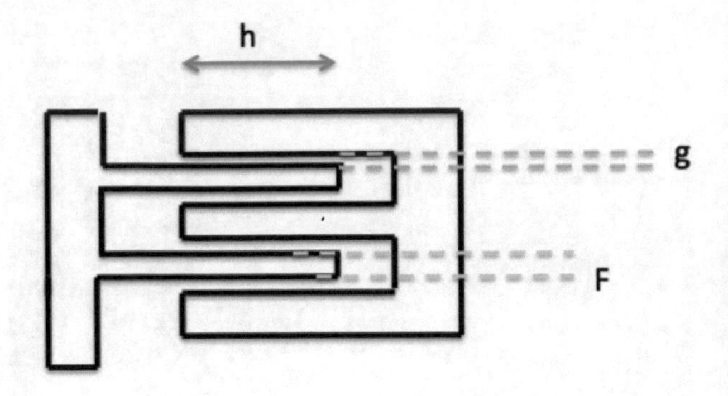

Fig. 7.9 Parallel plate—variable overlap

The obtainable power density depends on the size of the generator, the test mass, and the operating condition.

As an example, we can consider Fig. 7.9 as design corresponding to a parallel plate—variable overlap.

The parameter g describes the fixed distance between the surfaces of the movable and the fixed electrode. The capacitance $C(x)$ in function of the displacement x is described by the following equation:

$$C(x) = 2\epsilon h \frac{x_{\max} + x}{g} \tag{7.3}$$

where h is the height of the electrode. A similar procedure can be repeated for the other design.

References

1. Park, J., & Mackqy, S. (2003). *Practical data acquisition for instrumentation and system control*. Elsevier, Oxford.
2. Lacanette, K. (2003). National temperature sensors handbook. Handbook Ann. Mat. National semiconcductor
3. National Instruments (1996). *Data Acquisition Fundamentals, Application Note 007* (1996).
4. National Instruments (1996). *Signal conditioning fundamentals for PC-based data acquisition systems*, Handbook National Instruments.
5. Taylor, J. (1986). *Computer-based data acquisition system*. Instrument Society of America, USA.
6. Di Paolo Emilio, M. (2013). *Data Acquisition System, from fundamentals to applied design*. New York: Springer.
7. Roundy, S., Wright, P., & Pister, K. (2002). Micro-electrostatic vibration-to-electricity converters. In *Proceedings of ASME International Mechanical Engineering Congress and Exposition IMECE2002* (Vol. 220, pp. 17–22).
8. Stordeur, M., & Stark, I. (1997). Low power thermoelectric generator: Self-sufficient energy supply for micro systems. In *Proceedings of the 16th International Conference on Thermoelectrics* (pp. 575–577).

9. Shenck, N., & Paradiso, J. (2001). Energy scavenging with shoe-mounted piezoelectrics. *Micro IEEE, 21*(3), 30–42.
10. Roundy, S. (2003). *Energy Scavenging for Wireless Sensor Nodes with a Focus on Vibration to Electricity Conversion*. PhD thesis, University of California.
11. Tsutsumino, T., Suzuki, Y., Kasagi, N., Kashiwagi, K., & Morizawa, Y. (2006). Micro seismic electret generator for energy harvesting. In *Technical Digest PowerMEMS 2006*, Berkeley, USA, November 2006 (pp. 133–136).
12. Sterken, T., Altena, G., Fiorini, P., & Puers, R. (2007). Characterisation of an electrostatic vibration harvester. In *DTIP of MEMS and MOEMS*, Stresa, Italy, April 2007.
13. Sterken, T., Baert, K., Puers, R., & Borghs, S. (2002) Power extraction from ambient vibration. In *Proceedings of the SeSens*, Workshop on Semiconductor Sensors, Veldhoven, Netherlands, November 2002 (pp. 680–683).
14. Szarka, G., Stark, B., & Burrow, S. (2012). Review of power management for energy harvesting systems. *IEEE Transactions on Power Electronics, 27*(2), 803–815. ISSN: 0885-8993.
15. Cammarano, A., Burrow, S. G., Barton, D. A. W., Carrella, A., & Clare, L. R. (2010). Tuning a resonant energy harvester using a generalized electrical load. *Smart Materials and Structures, 19*, 055003.
16. Guyomar, D., Badel, A., Lefeuvre, E., & Richard, C. (2005). Toward energy harvesting using active materials and conversion improvement by nonlinear processing. *IEEE Transactions on Ultrasonics, Ferroelectrics, and Frequency Control, 52*, 584–595.
17. Mitcheson, P. D., Stoianov, I., & Yeatman, E. M. (2012). Power-extraction circuits for piezoelectric energy harvesters in miniature and low-power applications. *IEEE Transactions on Power Electronics, 27*, 4514–4529.
18. Szarka, G. D., Burrow, S. G., & Stark, B. H. (2012). Ultra-low power, fully-autonomous boost rectifier for electro-magnetic energy harvesters. *IEEE Transactions on Power Electronics, 28*(7), 3353–3362. doi:10.1109/TPEL.2012.2219594.
19. Maurath, D., Becker, P. F., Spreeman, D., Manoli, Y. (2012). Efficient energy harvesting with electromagnetic energy transducers using active low-voltage. *IEEE Journal of Solid-State Circuits, 47*(6), 1369–1380.
20. Beeby, S. P., Tudor, M. J., & White, N. M. (2006). Energy harvesting vibration sources for microsystems applications. *Measurement Science and Technology, 17*, R175–R195.
21. Khaligh, A., Zeng, P., & Zheng, C. (2010). Kinetic energy harvesting using piezoelectric and electromagnetic technologies – state of the art. *IEEE Transactions on Industrial Electronics, 57*(3), 850–860.
22. Paulo, J., & Gaspar, P. D. (2010). Review and future trend of energy harvesting methods for portable medical devices. *Proceedings of the World Congress on Engineering* (Vol. 2).
23. Zhu, D., Tudor, M. J., & Beeby, S. P. (2010). Strategies for increasing the operating frequency range of vibration energy harvesters: A review. *Measurement Science and Technology, 21*, 022001-1–022001-29.
24. Cepnik, C., Lausecker, R., & Wallrabe, U. (2013). Review on electrodynamic energy harvesters – a classification approach. *Micromachines, 4*(2), 168–196. http://www.mdpi.com/2072-666X/4/2/168. Accessed 20 January 2015.
25. Ulaby, F. T., Michielssen, E., & Ravaioli, U. (2010). *Fundamentals of applied electromagnetics* (6th ed.). Prentice Hall, USA.
26. Roundy, S., Wright, P. K., & Rabaey, J. M. (2003). A study of low level vibrations as a power source for wireless sensor nodes. *Computer Communications, 26*(11), 1131–1144.
27. Sazonov, E., Li, H., Curry, D., & Pillay, P. (2009). Self-powered sensors for monitoring of highway bridges. *IEEE Sensors Journal, 9*, 1422–1429.
28. Toh, T. T., Mitcheson, P. D., Holmes, A. S., & Yeatman, E. M. (2008). A continuously rotating energy harvester with maximum power point tracking. *Journal of Micromechanics and Microengineering, 18*, 104008-1-7.
29. Howey, D. A., Bansal, A., & Holmes, A. S. (2011). Design and performance of a centimetre-scale shrouded wind turbine for energy harvesting. *Smart Materials and Structures, 20*, 085021.

30. Razavi, B. (2002). *Design of analog CMOS integrated circuits*. New York: McGraw-Hill.
31. Razavi, B. (2008). *Fundamentals of microelectronics*. London: Wiley.
32. Sedra, A. S., & Smith, K. C. (2013). *Microelectronic circuits*. Oxford: Oxford University.
33. Razavi, B. (2002). *Design of integrated circuits for optical communications*. New York: McGraw-Hill.
34. Hurst, P. J. (2001). *Analysis and design of analog integrated circuits*. John Wiley & Sons.
35. Spies, P. (2015). *Handbook of energy harvesting power supplies and applications*. CRC Press book, France.

Chapter 8
Powering Microsystem

8.1 Power Conditioning

The power conditioning circuits play an essential role in an energy harvesting system through various parameters such as the input impedance, at the same time carries out processing functions such as power control and filtering. Advanced techniques actively influence the behavior of harvesting devices, such as piezo pre-biasing. The power limit of a system to be used was considerably reduced with conditioning circuits that operate at lower levels of power, by reducing the losses to increase the maximum efficiency of the harvesting system. The challenge is always to optimize the energy and the associated conditioning circuits to cope with a system where the correspondence of the power profiles and operation dynamics hours is in some way optimized. Power supplies are often intermittent and the excitation parameters may change over time. The studies give results measured on various sources of vibrations and is possible to see how they differ from those ideals. The purpose of the conditioning circuit is to avoid an oversized design, with a storage system for providing a correspondence between the temporal profiles of the power demand from the load source. In addition to the source of excitation, the impedance is a conditioning factor that determines the operating conditions of the system. While in general the source is not controllable, the input impedance is the main control mechanism. The input impedance of the conditioning system can be formed by a real part and an imaginary (resistive and reactive component), and it is synthesized by the action of active and passive components of the converter, usually controlled by the duty cycle of the active part and associated in the regulation circuits. Whereas the harvesting system can be modeled by a combination of linear and non-linear circuit elements, the maximum transmissible power is when the load is the complex conjugate of the output impedance. However, it is not always possible to work on this theory as a physical constraint, the voltage limitation or excursions can precluded [1–15]. Furthermore, the energy consumption associated with the resistive synthesizer and impedances of reactive load can become significant,

M. Di Paolo Emilio, *Microelectronic Circuit Design for Energy Harvesting Systems*, DOI 10.1007/978-3-319-47587-5_8

especially at low-power levels by causing a situation in which a different load can produce more power. To understand the behavior of a circuit, it is often useful to consider the waveforms of current and voltage in the frequency domain. For a resistive component, voltage and current appearing in phase and represent the power dissipated; in the case of phase out components, the fundamentals frequency components appear as reactive components and are described for the displacement factor, and represent the energy circulates between the source and the load that can modify the frequency response of the system; the non-linearity of the input impedance produces harmonic components that are described through the distortion power factor. They modulate the flow of energy between the source and the load to their particular frequency, the effect of which may or may not be significant in terms of average power, by depending on the Q-factor of the mechanical system. The impedance of the input resistive can be approximated by various types of circuits in a discontinuous conduction mode or by configuring a feedback loop around a converter to force the voltage to follow the current, or vice versa. An example is shown in Fig. 8.1 where a circuit layout consists of a rectifier and a boost converter.

The rectifier has not a store capacitor and the mosfet device is activated by depending on the input clock at various frequencies and duty cycles with the inductor in discontinuous conduction, in this case the inductor current falls to zero in every switching cycle. During switch on periods, the current ramp input can reach a level determined by the input voltage, then the average current (Idc_{ave}) follows the input voltage. In the discontinuous conduction mode, the input resistance can be approximated by the following equation:

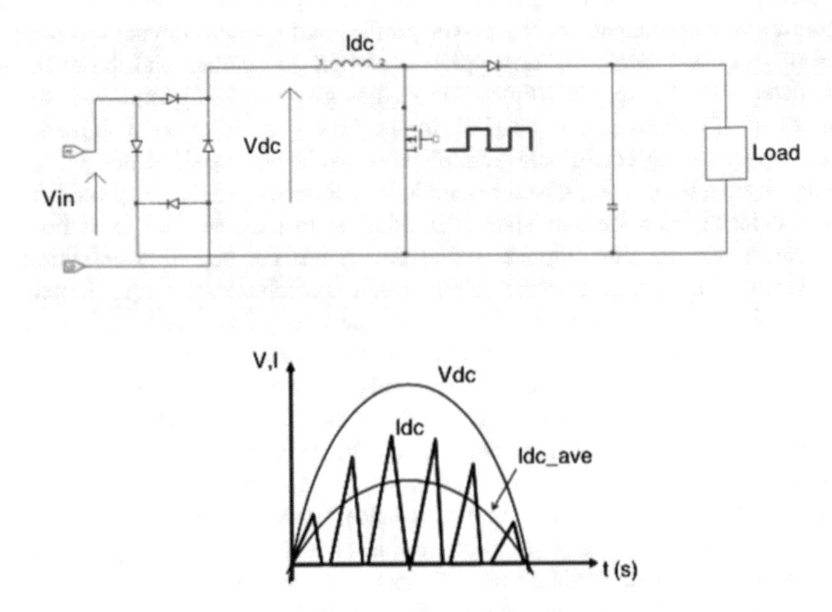

Fig. 8.1 Resistive load with discontinuous mode converter

$$\frac{V_{\text{out}}}{\Delta R_T} = -\frac{R_1}{(R_1 + R_0)^2} V_{\text{in}} \quad (3.11)$$

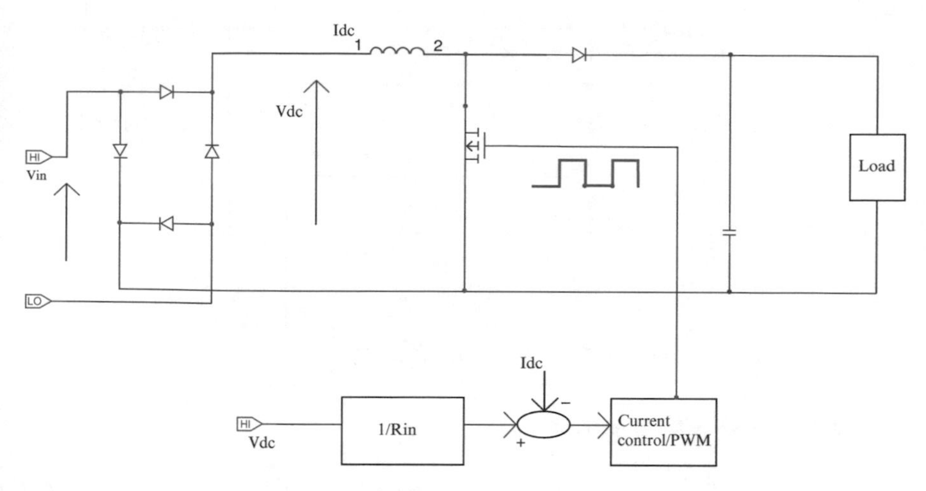

Fig. 8.2 Feedback control to force the phase of the current and voltage signals

$$R_{\text{in}} = \frac{2L}{DT} \tag{8.1}$$

where D and T are the duty cycle and the period of the switching waveform, respectively, and L is the inductor value. Another approach to obtain an impedance of resistive input is to employ a feedback to force the phase of the waveforms of voltage and current; this technique is widely used for the correction of the power factor in small-size power supplies. A block diagram of example system is shown in Fig. 8.2.

The circuit is in current control mode, through a negative feedback on the PWM controller in order to set the input current to the reference level. The reference current is derived from the waveform of the input voltage: i.e., the instantaneous voltage divided the desired input resistance R_{in}. In general, this circuit is difficult to employ it for very low levels of micropowering system. However, an important advantage is the operation of the converter with an inductor current in continuous or discontinuous mode. The power conditioning circuits with some element of the reactive input impedance determine the circulating energy between the power electronics and the harvesting of energy at the fundamental frequency. Resonators circuit can work in the "frequency tuning mode" to compensate for the reactive part of the input block. In many situations occur when the transducer is described by a mechanical energy system, common in piezoelectric devices. The linear circuits known as rotators are able to synthesize the reactive impedance of input and have found application in the suppression of the vibrations. However, they have a significant consumption of energy to synthesize the reactive components. For the energy harvesting, it is therefore necessary to base the layout around the switching circuits, and an example is shown in Fig. 8.3. Since a reactive load implies a flow of energy from the load to the source and vice versa, it is necessary to employ a

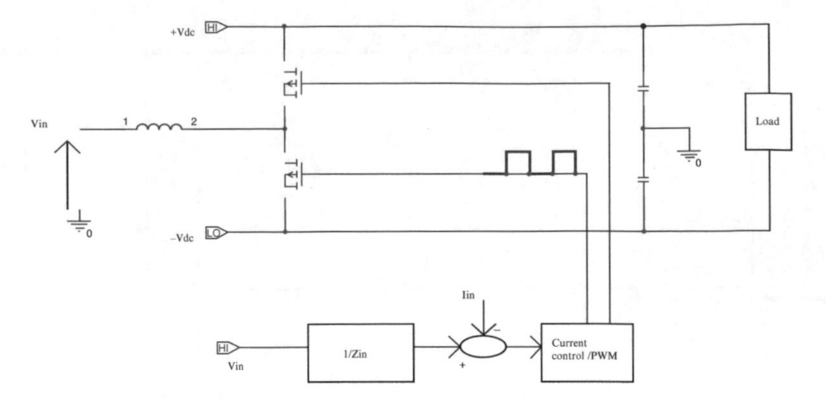

Fig. 8.3 Layout with reactive impedance

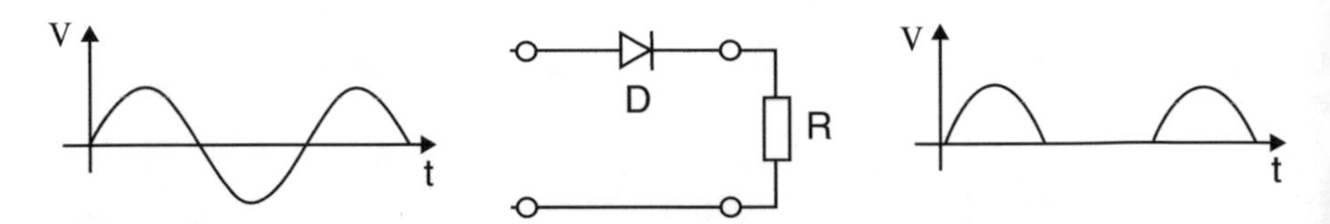

Fig. 8.4 General layout of a half-wave rectifier

converter that can operate in four quadrants. The converter operates in the average current control mode with the reference current derived from the input voltage.

8.2 Rectifier Circuit

The diode rectifiers circuits are widely used in electronic design, in particular in power supply and demodulation systems. The main objective is the signal conversion AC to DC that can be made with different circuit configurations, each with their advantages and disadvantages, used in various industries. The rectifier circuits are classified into two main groups, namely single-phase and three-phase. These stages are very important for industrial applications and for the transmission of energy into direct current (HVDC). The half-wave rectifier circuit is the simplest form as visualized in Fig. 8.4. For the majority of power applications, this type of circuit is not sufficient because the harmonic content of the rectifier's output waveform is very large and consequently difficult to filter. However, it is a very simple way to reduce the power to a resistive load (Fig. 8.4).

The full-wave rectifier circuit (Fig. 8.5), however, uses full waveform. This makes the rectifier most effective, and there is conduction on both half-wave on the cycle of the sinusoid, the smoothing (literally "spread the signal," or transform the AC signal into DC) becomes much easier and more effective. The classical

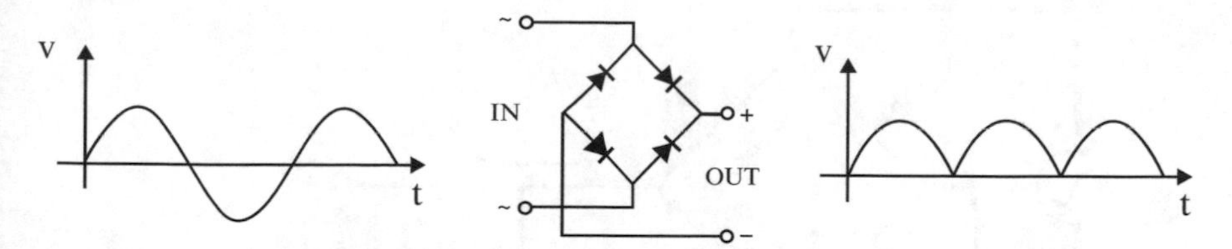

Fig. 8.5 General layout of a full-wave rectifier

circuit is based on four diodes in a bridge topology. The diodes can be replaced with active elements to provide the switching and increase the efficiency. The choice depends on the application of the circuit. While the full-wave circuits are mostly used with the bridge configuration, half-wave circuits may offer a better solution in some circumstances. For power applications are normally used power Schottky diodes which require only a forward voltage of about 0.2–0.3 V.

8.2.1 Bridge Rectifier Circuit

The bridge rectifier (single phase) uses four rectifier diodes, each connected in a closed-ring configuration ("bridge") to produce the desired output. The main advantage of this bridge circuit is that it requires a dual transformer. The diodes (D1–D4 in Fig. 8.6) are arranged in series, only two diodes conduct current during each half cycle of the sinusoidal waveform. During the positive half cycle, diodes D1 and D2 are in conduction, while the diodes D3 and D4 are reverse biased and current flows through the load as shown in Fig. 8.6. Conversely, during the negative half cycle.

Bridge rectifiers integrated components are available in a range of different voltages and dimensions that can be welded directly into a PCB circuit (Fig. 8.7). Depending on the technology of the diode threshold voltage, each may vary in the neighborhood of 0.6 V.

For a complete description, several parameters must to be considered: for example, repetitive peak voltage (VRRM) and the Reverse Recovery Time. The repetitive peak voltage (VRRM) is the actual value of maximum allowable reverse voltage across the diode rectifier.

When switching from the conducting to the blocking state, a rectifier has stored charge that must first be discharged before the diode blocks reverse current. This discharge takes a finite amount of time known as the Reverse Recovery Time (t_{rr}).

The smoothing capacitor placed in the output converts the wave into a DC voltage. Generally for power circuits in current continues, the smoothing capacitor is an electrolytic that has a capacitance value of the order of 100 μF or more. However, two parameters are important to consider when choosing a smoothing capacitor, these are its operating voltage, which must be greater than the value of load output

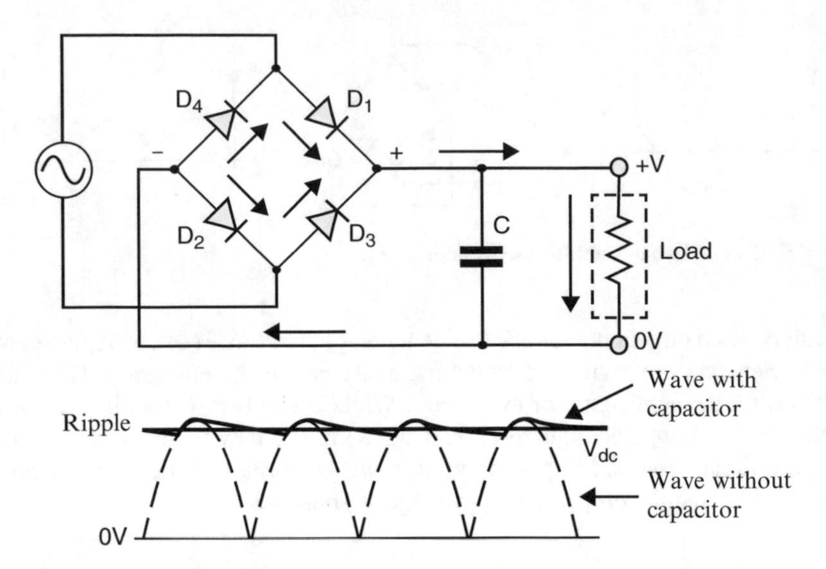

Fig. 8.6 Example of bridge rectifier circuit

of the rectifier, and its capacitance value, which determines the amount of ripple (variable part of the continuous signal) which will be superimposed on the DC voltage. As a general rule, you can think of a ripple voltage of less than 100 mV peak to peak. The amount of ripple voltage can be virtually eliminated by adding a filter to the output terminals of the rectifier bridge. This low-pass filter is constituted by two capacitors, generally of the same value, and an inductance in order to induce a path of high impedance for the alternating ripple component. Another alternative layout most practical and economical is to use a voltage regulator, such as an LM78xx (where "xx" stands for the output voltage) that can reduce the ripple of over 70 dB offering at the same time a constant output current of more than 1 A.

An important aspect of the rectifier circuits is the loss in the output voltage, caused by the voltage drop of the diodes (about 0.6 V for silicon and 0.3 V for the Schottky diodes). This reduces the output voltage limiting, therefore, that available output. The loss of voltage is very important for low-voltage rectifiers (for example, 12 V or less), but is insignificant in high voltage applications such as HVDC. The rectifiers are also used for the detection of the amplitude modulated signal relatively to radio signals. Another typical use of rectifier circuits is the power supply design. A power supply can be divided into a series of blocks (Figs. 8.8 and 8.9), each of which performs a particular function: a transformer and a rectifier circuit for converting the AC signal into DC. The power supplies are designed to produce less ripple that can cause several problems. For example, in audio amplifiers, too much ripple looks like an annoying buzzing noise; in video circuitry, an excessive ripple causes defects in the image; in digital circuits can cause erroneous results of logic circuits.

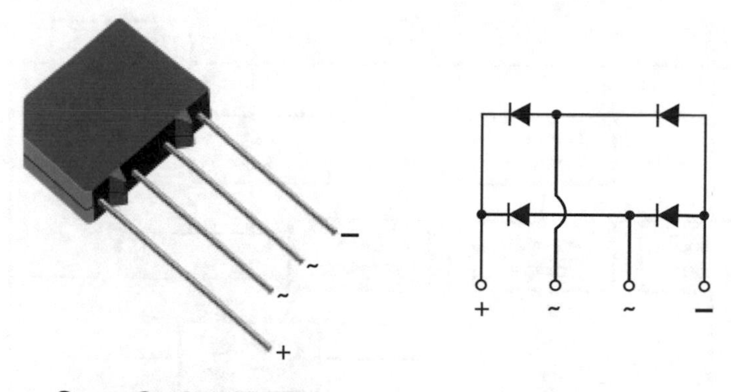

PRIMARY CHARACTERISTICS	
Package	KBPM
$I_{F(AV)}$	2.0 A
V_{RRM}	50 Vto 1000V
I_{FSM}	60 A
I_R	5 μA
V_F	1.1 V
T_J max	165°C
Diode variations	In-Line

Fig. 8.7 Example of diode bridge of Vishay

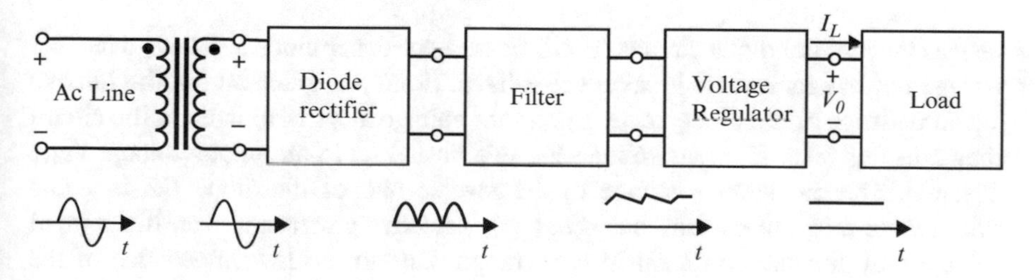

Fig. 8.8 Block diagram of a power supply

8.2.2 Zener Diode as Voltage Regulator

The zener diodes are widely used in reverse bias to produce a stabilized output voltage. When connected in parallel to a variable voltage generator in reverse bias,

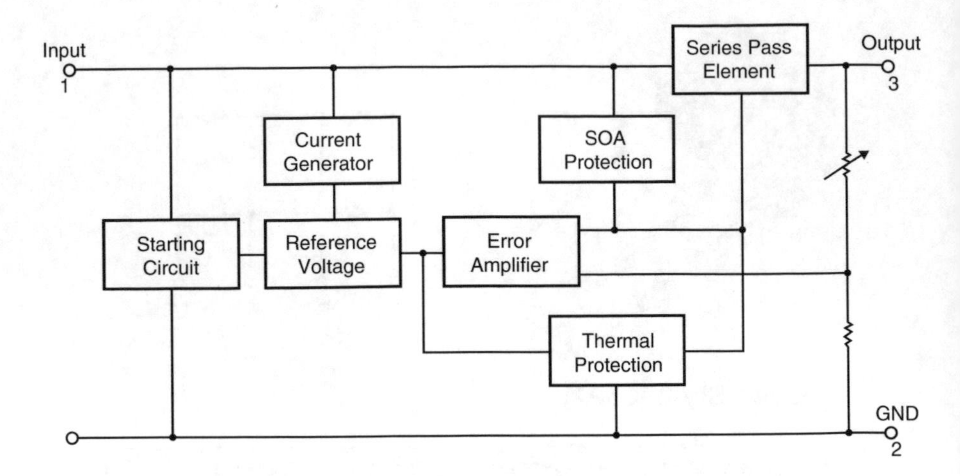

Fig. 8.9 General diagram of power supply

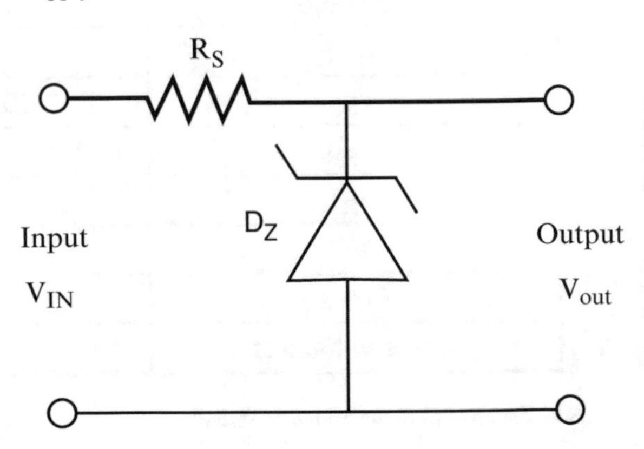

Fig. 8.10 General diagram of a voltage regulator with zener diode

such as the rectifier diode circuits just discussed, the zener diode conducts when the voltage reaches its reverse breakdown voltage. From that moment on, the relative low impedance of the diode keeps a constant voltage at its terminals. In the circuit shown in Fig. 8.10, an input voltage V_{IN} is adjusted up to an output voltage V_{OUT} is stable. The breakdown voltage of the reverse bias of the diode *DZ* is stable over a wide range of current and keeps V_{OUT} relatively constant even if the input voltage may fluctuate on a fairly wide range. Due to the low impedance of the diode, the resistor *RS* is used to limit the current through the circuit. The value of the resistance must satisfy two conditions: it must be small enough so that the current through the zener diode is sufficient to maintain the conditions of reverse breakdown; moreover, it must also be large enough for the current through *DZ*. A small problem with stabilizer zener diodes circuits is the presence of electrical noise in the attempt to stabilize the voltage. Normally this is not a problem for most applications, but the addition of a decoupling capacitor of large capacitance through the output of the zener may be needed to obtain a further stabilization of the voltage.

8.2.3 Considerations

The rectifier circuit is perhaps one of the most well-known circuits, formed by diodes and capacitors. It's widely used as the first stage in the conditioning circuit in energy harvesting applications. However, the simple pattern as shown in the preceding figures implicates some considerations. The classic circuit in Fig. 8.11 leads us to consider the R_s values, the classic case is having R_s tiny order of some ohms, by resulting in a waveform shown in Fig. 8.12.

Generally, an impedance of the source will be comparable with that of the load, then waveforms close to those of Fig. 8.12b. By a control of the waveforms, it can be noted that the energy flow is directional uniform in peak rectifier. Namely, the peak rectifier behaves as a non-linear resistive load. The peak rectifier is a simple way to synthesize the impedance of the resistive input, by providing an important adjustment function, which can be scaled to a wide range of power levels; but the current harmonics may prove of falling in many of powering systems. However,

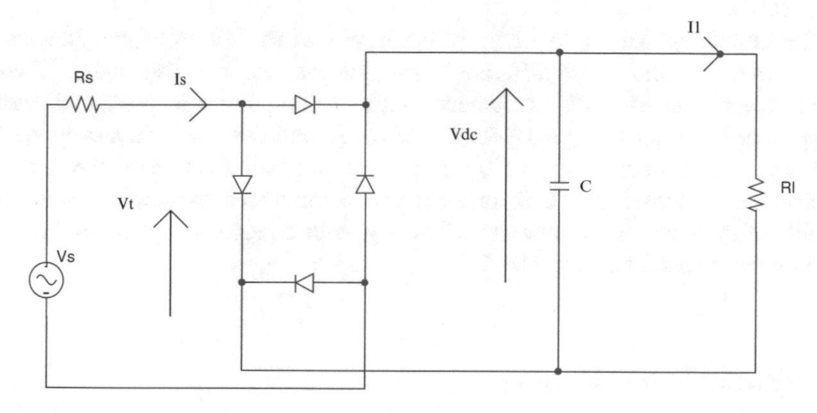

Fig. 8.11 Classic rectifier circuit

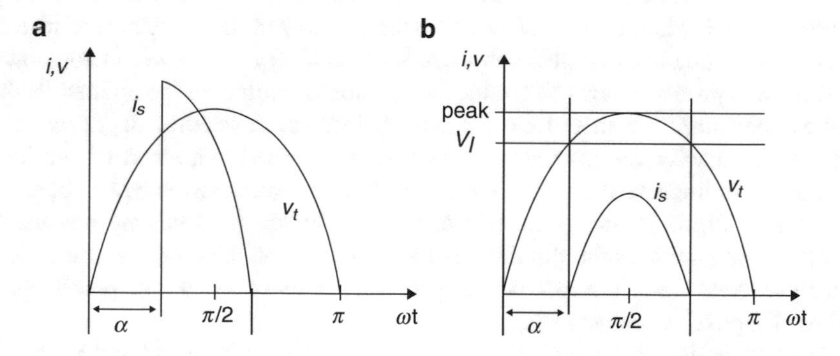

Fig. 8.12 Waveforms with (**a**) the load source impedance much smaller, and (**b**) the source impedance close to that of the load

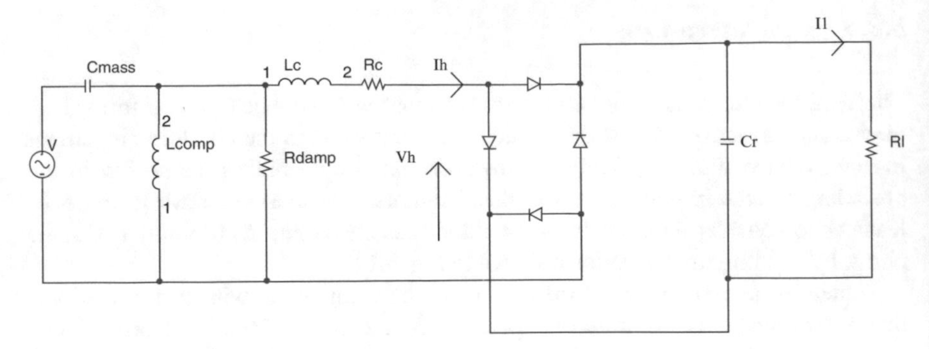

Fig. 8.13 Vibration energy harvester

in some harvesting applications, the electronic characteristics minimize the impact of current harmonics. In Fig. 8.13 is shown the equivalent circuit of a simplified vibration energy harvesting resonant circuit with electromagnetic transduction and peak rectifier.

The circuit appears as a filter by the time that the different components of the rectifier harmonics see different output impedances, and especially different resistive components. The fundamental current component is presented with a component of resistive origin of $R_{coil} + R_{damp}$. A well-designed harvesting system will have a resistance of the coil, which is some small fraction of the damping mechanical. By introducing a distortion power factor, it enables simple conditioning circuits of systems able to operate efficiently, with a half-wave peak rectifier that produces more useful energy [16–25].

8.3 Piezoelectric Biasing

The energy harvesting systems that use piezoelectric transduction have output impedance which is highly reactive due to the capacity of the piezoelectric material and therefore the ideal complex conjugate load has a very low-power factor. Since a typical piezo produces few milliwatts, the attempt to synthesize the optimal loading with a circuit similar to that of Fig. 8.3 is likely to cause a system with greater power generation capacity; so designers IC systems have turned to a series of non-linear approaches to improve the power without large overhead costs. A typical approach is shown in Fig. 8.14 and is called synchronous switched harvesting on inductor (SSHI), which involves the flipping of charge polarity on the piezoelectric material twice per cycle when the mechanical part reaches its maximum displacement by means of a physical inductance.

Another biasing approach is shown in Fig. 8.15. The switches S1 and S4, as well as S2 and S3, operate in pairs. When the piezoelectric material reaches its maximum deflection, one of the pairs of switches is turned on, discharging the energy from

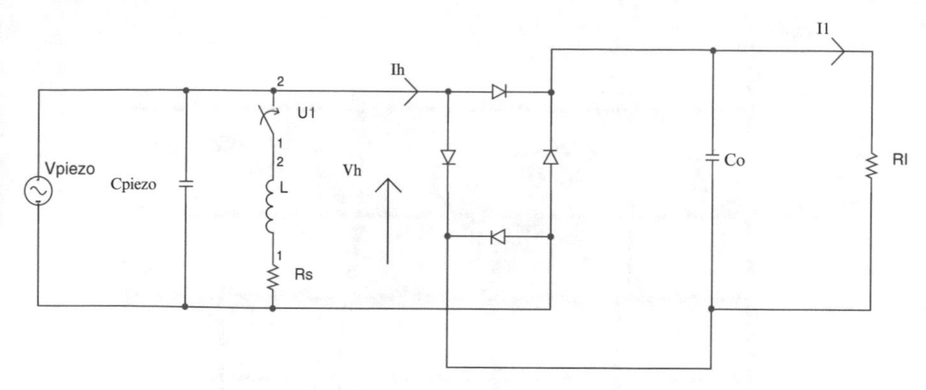

Fig. 8.14 Circuit of synchronous switched harvesting on inductor (SSHI) with rectifier

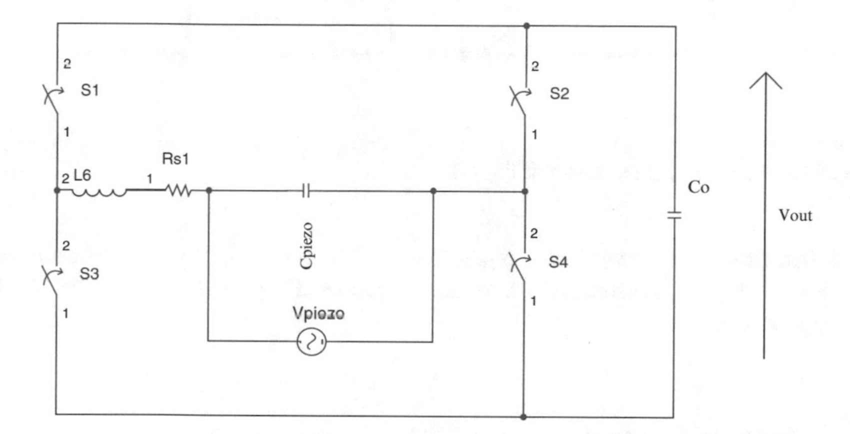

Fig. 8.15 Single-supply pre-biasing for piezoelectric

the piezoelectric capacitance in the capacitor through the series inductor (discharge phase in Fig. 8.16), all very quickly with respect to the mechanical excitation frequency of the system. As soon as the discharge phase is completed, the voltage on the piezoelectric capacitor reaches zero and the opposite switch pair actives and injects a certain piezoelectric charge on the capacitor of opposite polarity. So it increases the force with which the transducer is able to oppose the relative movement between the mass and the base, thereby by increasing the electrical damping. The piezoelectric material is then moved to its extreme opposite, thereby by increasing the voltage. The process is repeated with the generation of the output voltage. The piezoelectric capacitor remains in open circuit during the entire movement of the beam, while in SSHI the piezoelectric element is short-circuited by the switching of the rectifier diode. To extract energy from a transducer, the force produced by the transducer which opposes the motion must be in phase with the speed of the transducer. Due to the impedance of the output capacitive, a piezoelectric transducer to open circuit has strength and speed 90° out of phase

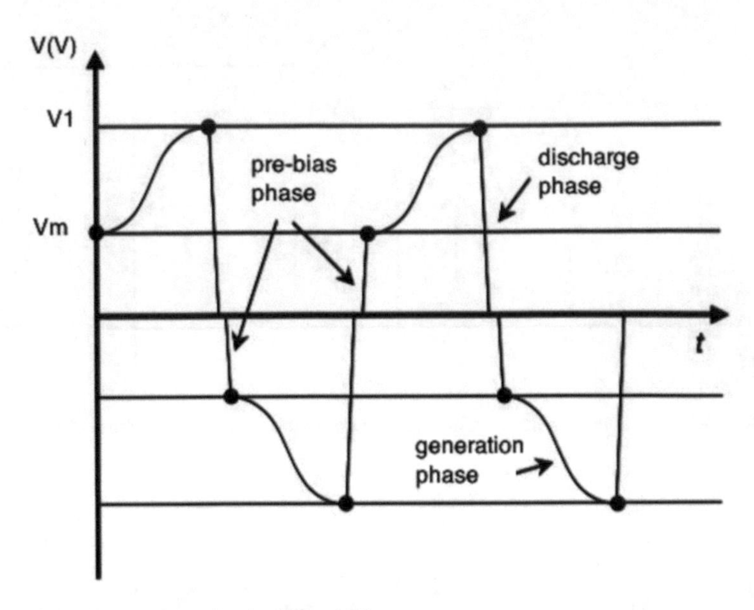

Fig. 8.16 Waveforms of the circuit of Fig. 8.15

and therefore has a power factor equal to zero. The discontinuity introduced in the waveform (Fig. 8.16) effectively corrects the phase of the fundamental with nearly unity power factor.

8.4 Voltage Control

The output voltage adjustment process is to maintain constant value in function of load variations. The switching converters are power converters that allow to keep the output power without losses, and then adjust the switched output typically modify the input power supply by altering the impedance, usually through an automatic feedback loop. The voltage control systems require a monotonic relationship between the power and the impedance over the operating range of input. To increase the power with a voltage source, the load resistance should be reduced; to increase the power with a power generator, the load resistance must be increased. The power response of an energy harvesting source has in general not only the regions in which the resistance load increases the power, but also the regions in which the opposite is true, in this way a simple feedback control is not possible when it operates near the peak power region. A similar problem occurs with photovoltaic panels. Because of the difficulty in the use of the power converter to perform a feedback voltage control by controlling the impedance of the converter input, energy harvesting applications can benefit from the alternative voltage control strategies. A possible example is to separate the voltage regulation from that impedance

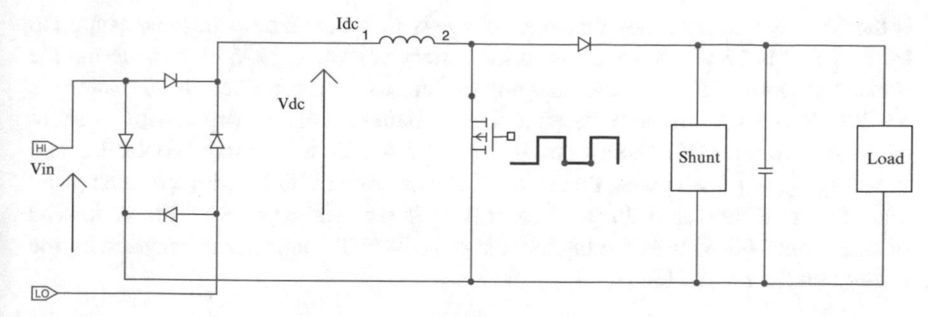

Fig. 8.17 Powering conditioning with a shunt regulator

input, for example, using a converter with fixed input impedance and by using a secondary shunt voltage regulator (Fig. 8.17). A possible layout is a flyback converter operating in discontinuous open loop mode with a shunt regulator and a low-power conditioning system [26–30].

8.5 MPPT

The power sources such as solar cells are commonly employed to work at a power point (peak) by using a linear feedback system to ensure constant peak operation. The type of layout is called MPPT also used for the energy harvesting. The power converters tend to perform small adjustments to their working point (input impedance), trying to find the point with the maximum output power or the point at which the derivative respect to the operating point is zero power. The power can be calculated from measured currents and voltages or deducible from other variables when the voltage change is slow and compared to the capacitor charging rate. When monitoring systems are used with the vibrations energy harvesting, it is important to take note of the energy stored in the mechanical oscillator. When the dynamic load is considered, it can be shown that the output power is a function of the load impedance and then of impedance changing rate. The MPPT controllers are in fact able to use all the power generated by the solar panel to charge the battery, unlike traditional PWM controllers that send to the battery the current generated by the panel. To understand this concept, we must first specify that the power of a panel is the result of the following multiplication: Current supplied from the panel for voltage generated by the panel. The working voltage generated by the panel is typically around 16–18 V (not 12 V, as the battery voltage): this voltage surplus is not considered in conventional voltage regulators, unlike the MMPT regulators. We assume that the current generated by a photovoltaic panel is, in a certain situation, 3 A: with a traditional PWM controller the current that is transferred to the battery for charging is equal to 3 A. An MPPT controller instead analyzes the power generated by the panel ($P = V \times I$), and therefore also considers the panel voltage:

if therefore we suppose that the panel voltage is 17 V at that time the power supplied by the panel is 17 V × 3 A = 51 W. If the battery voltage is 13 V, by considering the maximum power of 51 W, the charging current to be transmitted to the battery is 51 W/13 V = 3.9 A. We note therefore that the battery will be charged with a current of 3.9 A with the MPPT controller, instead of 3 A with a traditional controller, and the charging will take place, therefore, at a faster rate of 30 %, with the same panel and of current delivered. In practice, it is as if we used a panel of 130 W instead of one about 100 W, then the higher cost of an MPPT controller is balanced by the savings on the cost of the panel.

8.6 Architecture

In the conditioning system with passive circuits, such as the rectifier, the problems are concentrated on the input impedance: a zero voltage to the storage capacitor, the harvesting circuit looks like a short circuit. The problem is exacerbated if the capacitor has a relatively large capacity and can take many minutes to load. In addition to this it is necessary that during start-up the load does not require a lot of start-up energy; to alleviate this is necessary to enter in the circuit active component connections. The addition of active elements in the power conditioning to control the converters presents a further problem of power supply. If the output is AC harvesting system, a diode network can be used to provide power and start the main converter. A simplified block diagram of this simple design concept is shown in Fig. 8.18. The further advantage is the ability to initiate the harvester circuit when it has a peak output voltage less than that required to power the active circuitry. By incorporating a provision multiplier diode is possible to start when the harvester has a peak output voltage less than that required to power the active circuitry.

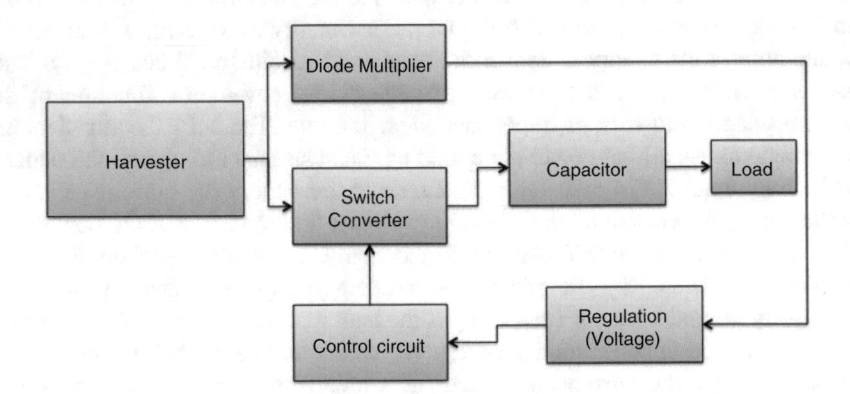

Fig. 8.18 Block diagram of a power conditioning circuit with parallel diodes for starting of the main converter

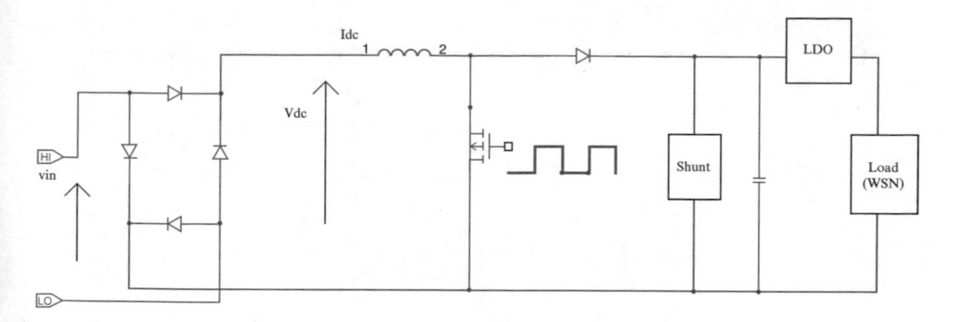

Fig. 8.19 Conditioning circuit with duty-cycled power demand

A further complication arises with the energy storage capacitor, to minimize the value of the capacity to promote the rapid start-up will result in greater fluctuations in voltage for a given load profile, and therefore requires a voltage regulation circuit. In Fig. 8.19 a circuit example with a voltage regulator is shown.

The energy is stored in a capacitor with shunt adjustment followed by a low-dropout regulator (LDO) by providing, for example, to the WSN load a constant 3.3 V. When switched on, the system takes a little more than 100 s to fully charge the capacitor (6.8 V). The load can then perform a transmission operation of 12 s. The capacity must be selected so that its voltage is to a minimum (3.3 V) at the end of the transmission period, by avoiding voltage variations. WSN then goes to low-power mode and the capacitor voltage recovers enough to get in a next transmission period after 81 s. The duty cycle of WSN must always remain below 15 %. A switching converter can be used in place of the linear regulator; however, it is not always the case that the circuital additional complexity justifies the efficiency gains.

8.7 DC-DC Systems

In energy harvesting applications, the DC-DC converters play an essential role to power electrical loads. The various energy transducers analyzed in the previous chapters provide various output voltages, which are typically too high or too low to power direct electrical loads. Therefore, the DC-DC converters have the objective of providing a stable supply voltage.

8.7.1 Linear Regulators

The task of linear regulators is to establish a constant output voltage regardless of the output current and input voltage. To check the adjustment element, an amplifier loop

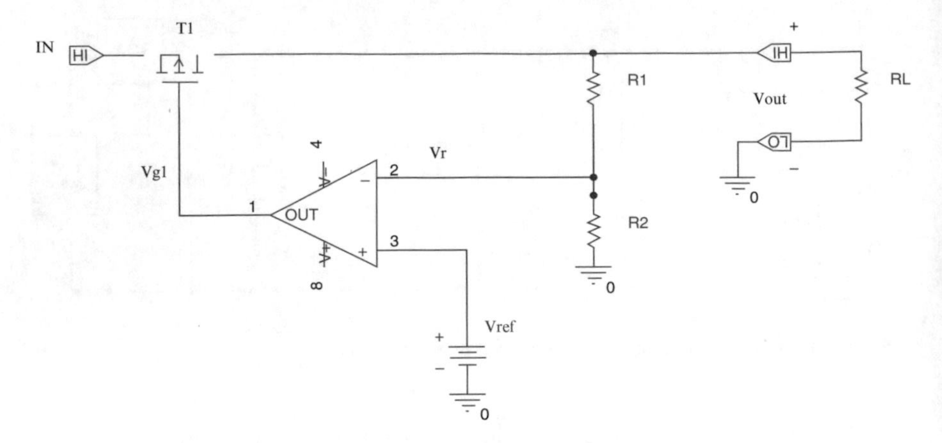

Fig. 8.20 Block diagram of a linear regulator

is used together with a more resistors network. The operational amplifier performs the task by amplifying the signal, and then compare it with a reference voltage. In Fig. 8.20 a classic example of the linear regulator is shown.

The loop system controls the voltage amplification that depends on the V_{ref} connected to the positive terminal of the operational amplifier. The disadvantage is the power loss due to the voltage drop between input and output voltage. This means that the efficiency is high when the difference between input and output voltage is low. Furthermore, a linear regulator cannot have an output voltage larger than the input. The advantage of this concept in comparison to the switching regulators is the absence of the switching elements. Therefore, there is no ripple in the generated output. In reference to Fig. 8.20 can be calculated the efficiency defined as:

$$\eta = \frac{P_{\text{out}}}{P_{\text{in}}} = \frac{V_{\text{out}} I_{\text{out}}}{V_{\text{in}} I_{\text{in}}} = \frac{V_{\text{out}}\left(I_{\text{in}} - \frac{V_{\text{ref}}}{R_2}\right)}{V_{\text{in}} I_{\text{in}}} = \frac{V_{\text{out}}}{V_{\text{in}}}\left(1 - \frac{V_{\text{ref}}}{I_{\text{in}} R_2}\right) \tag{8.2}$$

By referring to efficiency equation, the maximum possible value is obtained by minimizing the resistors R1 and R2. The efficiency depends on the ratio between the output and input voltage.

8.7.2 Switching Regulators

The switching regulators are an alternative to linear regulators. They also fulfill the task of generating a constant output voltage and independent from input voltage and output current. The regulation element acts as a switch that can be in the state on or off. The types of switching regulators can be divided into four main subtypes. A buck converter delivers a lower output voltage and a boost converter with an output voltage higher of the input voltage. These two types have in common an inductor as an energy storage element. For these two types of converter two different modes of operation are defined: in the continuous conduction mode (CCM), the current is always greater than zero, while in discontinuous conduction mode (DCM), the current is zero for a certain period of time. The flyback converter uses a transformer instead of an inductor and transfers the input energy during the off-state of the switch to the output capacitor. This type of converter is mainly used in powers of greater than 100 W output, for which it is not important state-of-the-art applications for energy harvesting. Another important type of converter is the charge pump: several switching transistors are used and only capacitors are used as storage elements. A less common type of converter is Meissner, useful for energy harvesting applications in which the energy transducer delivers output voltages less than 500 mV. As in flyback converters, a transformer is used but in this case the secondary winding is used to control the switching transistor.

8.7.3 Buck Converter

The step-down converter (buck) provides a lower output voltage of its input. Basically, the input voltage is periodically turned on and off and a low-pass filter transfers the average value of the converter output. The basic scheme of the converter is shown in Fig. 8.21. While a typical electric circuit of a buck converter is shown in Fig. 8.22. A voltage proportional to the output voltage is compared with a reference voltage according to the desired output voltage. The difference between these two signals (V_{err}) is amplified. The output of the comparator is a PWM signal of the T1 switching transistor.

8.7.4 Boost Converter

A step-up converter (boost) is used when in need a large voltage from an input source. Regarding energy harvesting transducers, it is an important device for thermogenerators and inductive generators, in which the output voltage is typically less than 1 V. In Fig. 8.23 is shown the physical principle, while in Fig. 8.24 an example of a circuit. In the first phase the switch is closed and the current of the inductor ideally increases with a linear profile, therefore, the energy stored in the inductor is increasing. In the second phase the switch is open and the energy stored in the inductor is transferred to the exit where the capacitor C can be connected to a second storage element. The variation of the current due to the open switch of the inductor induces a voltage which is added to the input voltage. In this way, the two phases are alternate periodically and the output voltage is always greater (or equal) than converter input voltage.

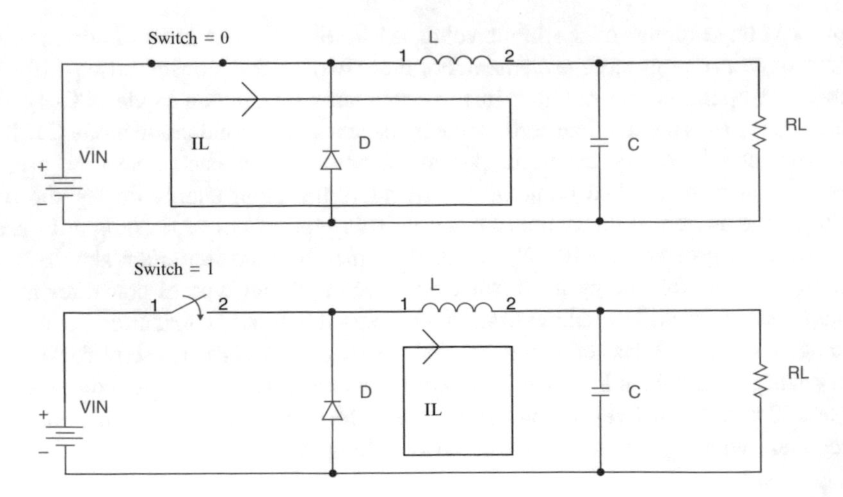

Fig. 8.21 Physical principle of a buck converter

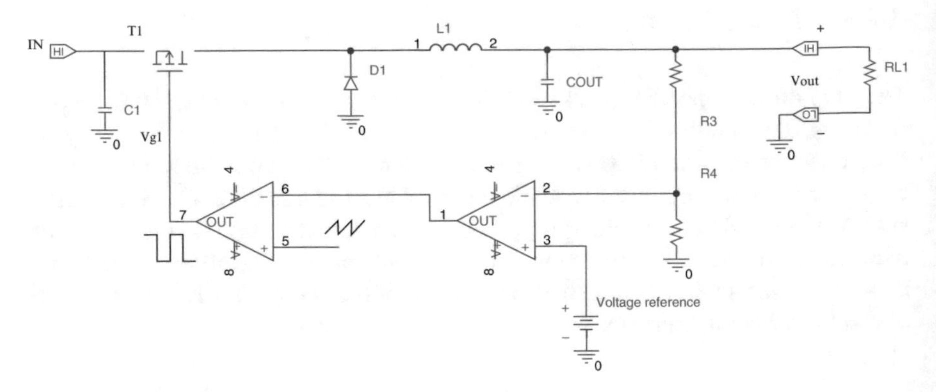

Fig. 8.22 Electrical layout of a buck converter

8.7.5 Buck-Boost Converter

The inverting converter is also called Buck-Boost because it refers to two types of converter, a slightly more complex, which provides for the union of the two circuits previously seen by using only one inductor in common with two switches and two diodes, while the second that we will analyze in this paragraph can be seen as a modification of one of the two. In this type of circuit two distinct meshes appear, one that includes the switch placed in series and the inductor in parallel, and the second created only by resistive and capacitive load. The diode is placed in series, but in the opposite polarity to the previous configurations and constitutes a link between the two circuital regions (Fig. 8.25).

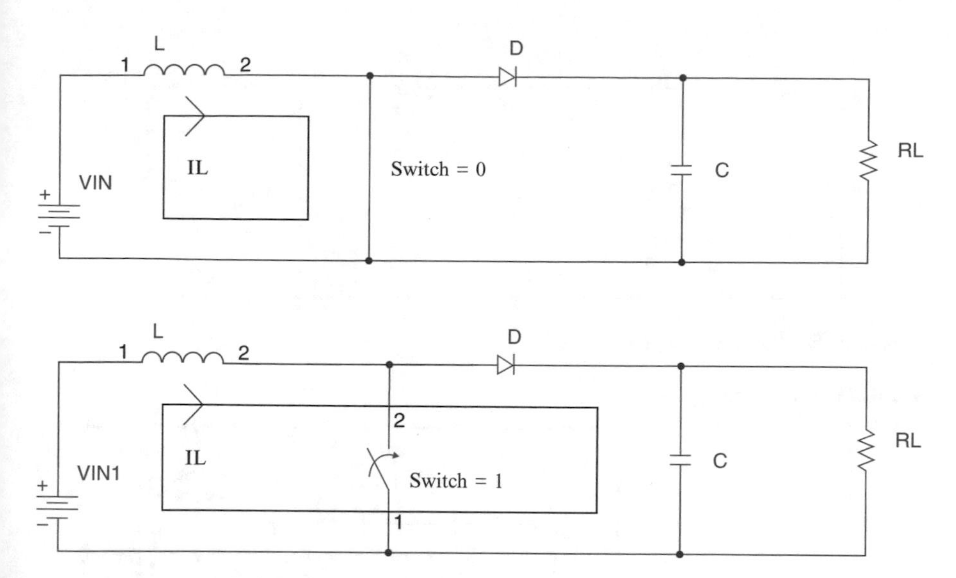

Fig. 8.23 Physical principle of the boost converter

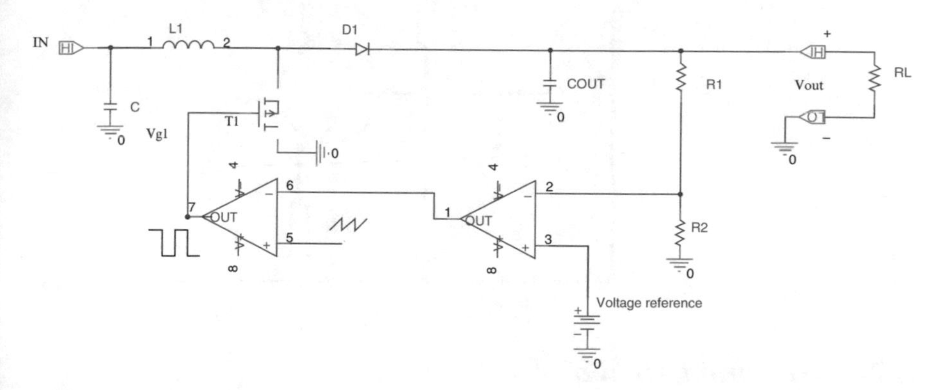

Fig. 8.24 Example of circuit layout of the boost converter

Even this circuit operates with two states of operation, of course scanned by the switch. To the initial conditions of course, all passive components will be discharged and free of stored energy, but when the power transistor is switched on, the current will be forced to flow from the diode on the inductor which stores magnetic energy. When the switch is opened, the inductor is forcing the current on the second mesh, by going to charge the capacitor and supplying the load. So, by returning to the switch closed again, the inductor will load again, while the capacitor will provide enough energy in the output having still stored electrical energy. Of course with this operating cycle the output current and thus the polarity of the voltage will be opposite to that of input; hence the name of inverting [2, 31–35].

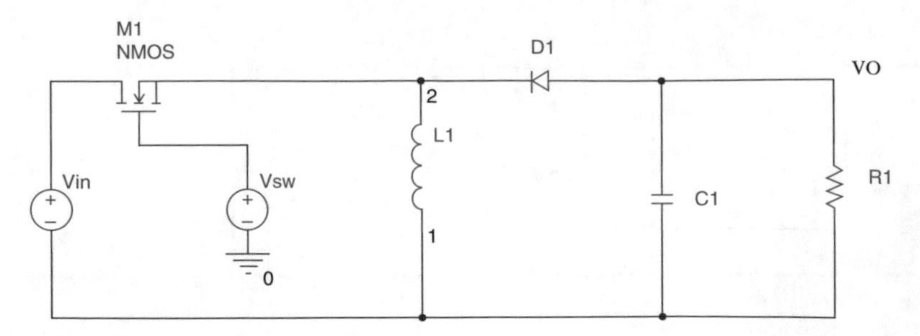

Fig. 8.25 Buck-boost converter

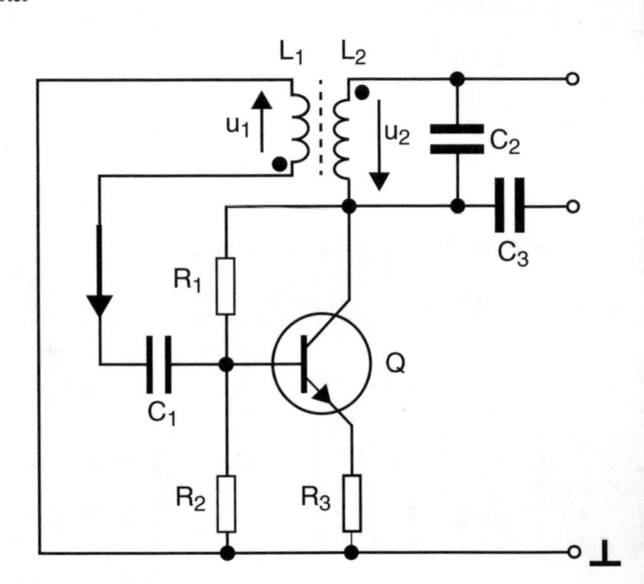

Fig. 8.26 Armstrong oscillator

8.7.6 Armstrong Oscillator

For the energy harvesting applications, it may be advantageous to use as the Meissner or Armstrong oscillator to build a switching converter. In this way it is created a self-oscillating circuit in which the switching transistor is driven by a winding of a transformer. Consequently, the adjustment circuit can be simplified, since it is not necessary any more in the clock circuit with a driver to control the switching transistor. This also helps to save energy in low load conditions (Fig. 8.26).

8.8 Load Matching

Since there are some energy transducers such as a thermoelectric generator (TEG) based on the Seebeck effect or a solar cells with a parasitic ohmic internal resistance, it is useful to think of the load matching in order to obtain the maximum power. A transducer is generally modeled as a real voltage generator. If the interface circuit for the transducer absorbs current, the measure will be affected by an error proportional to the voltage drop across the internal resistance of the generator, therefore is important that the acquisition circuit has high input impedance. It is also important that the first stage of the acquisition has low output impedance, so as to provide the acquired signal to downstream circuits regardless of the current absorbed by them (Figs. 8.27 and 8.28). These capabilities are defined impedance matching and are typically made with an operational amplifier in buffer configuration.

It can be important to realize a connection that makes the maximum voltage across the load or to maximize the current or power level. In other situations it may be important to minimize the deformation of the waveform on the load compared to that provided by the generator. Hence we can distinguish four mode of matching between the generator and load:

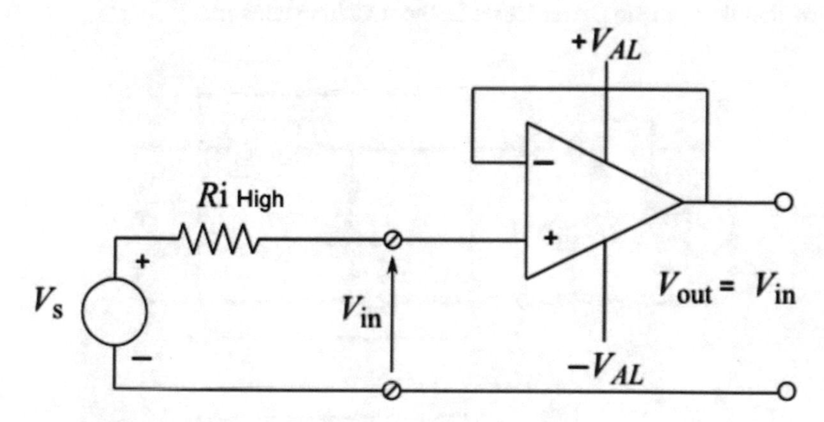

Fig. 8.27 Impedance matching

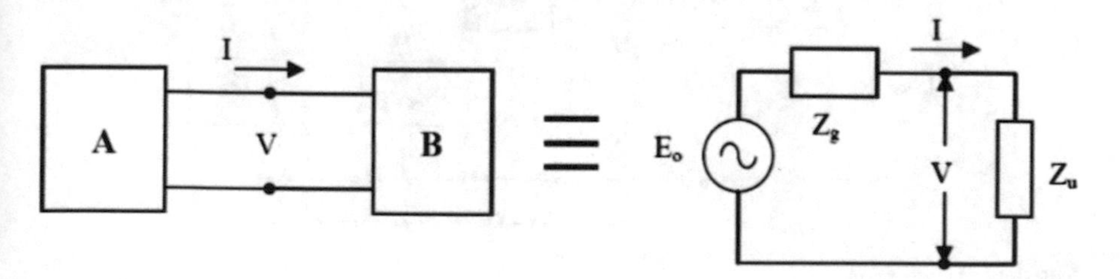

Fig. 8.28 General layout of matching

1. Maximum voltage matching
2. Overcurrent matching
3. Maximum power matching
4. Matching of uniformity (minimum waveform distortion)

The condition for which there are no signals of reflections is that the voltage on the load is in phase with that of the generator. Or if it is provided the following condition:

$$Z_u = Z_g \tag{8.3}$$

Taking into account some considerations, the energy matching condition results from the following condition:

$$Z_g = Z_u^* \tag{8.4}$$

The technique of "L matching" allows to do the matching of both real and imaginary part of that impedance Z_T of the transducer. It is implemented by using two reactive components between the source and the load as in Fig. 8.29. The reactive components may be inductors or capacitors and allow to adapt any load impedance Z_T to any generator impedance Z_G. The use of reactive elements also allows that there are no power losses in the matching network.

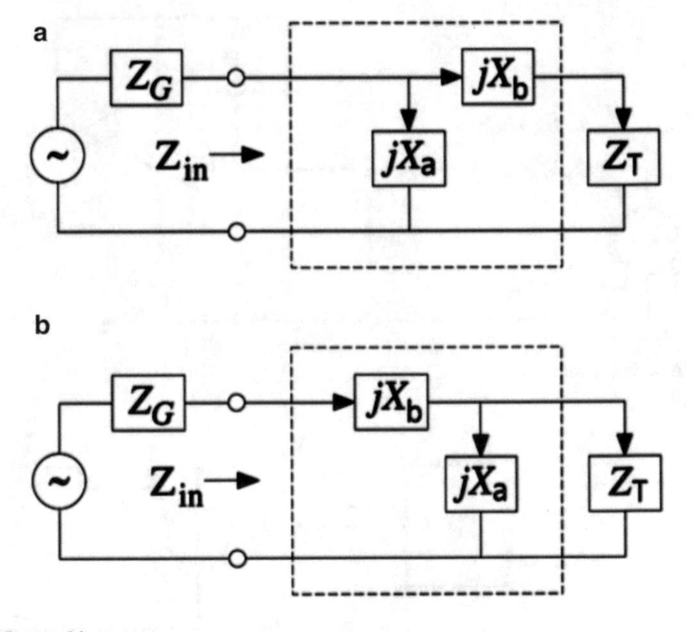

Fig. 8.29 L matching

8.9 AC-DC Systems

Vibrations are a ubiquitous environment source used in energy harvesting systems. Piezoelectric, electrostatic, and electrodynamic transducers provide AC power from ambient vibrations. Therefore, AC-DC converters are needed in order to convert their power into DC current which is required by the load of the energy harvesting system. The AC-DC converter used for such transducers is normally a two-stage power converter which consists of the following elements:

- AC-DC rectifier. The adjustment is usually done with a bridge rectifier of diodes or voltage multipliers for the case of electrodynamic transducers to increase the voltages. The current multipliers are used in piezoelectric transducers to increase the low output current and rectifying the output power.
- DC-DC converter. After the rectification of the alternating current, it is necessary to adapt the transducer voltage level to the load.

A rectifier is necessary when the piezoelectric transducers are employed in an energy harvesting system. Voltage or current rectifiers-multipliers are an alternative to full-wave rectifiers when is need to increase the voltage or current, respectively, and correct the signal. There are three different rectifiers AC-DC: diode-resistor pair rectifier, a diode-pair rectifier, and a synchronous rectifier (Figs. 8.30, 8.31, and 8.32). A charge pump is used to provide a regulated output voltage to a filter

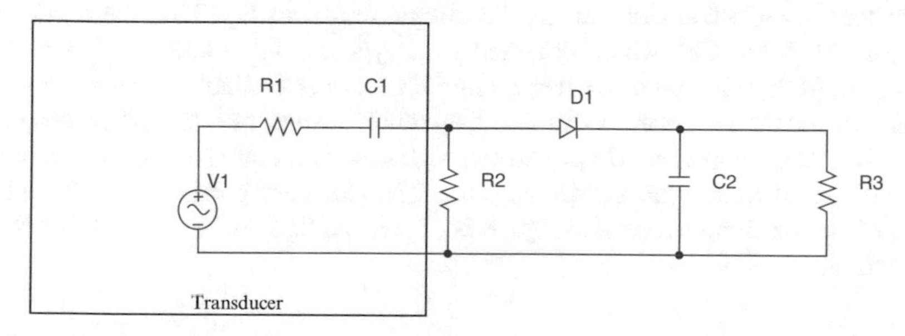

Fig. 8.30 Converters AC/DC, diode-resistor pair rectifier

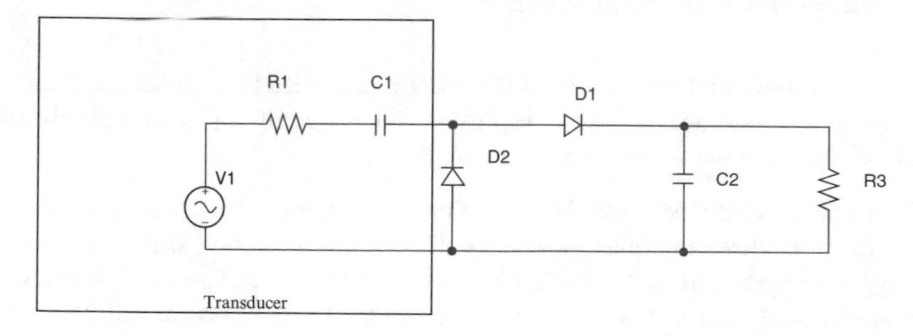

Fig. 8.31 Converters AC/DC, passive full-wave rectifier

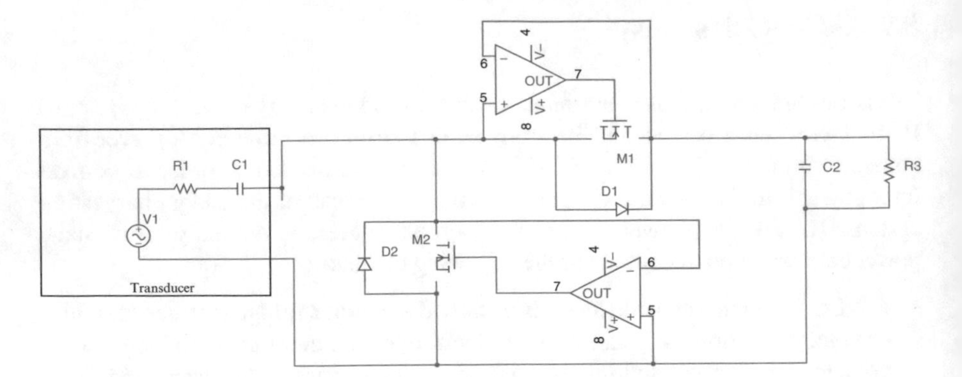

Fig. 8.32 Converters AC/DC, synchronous full-wave rectifier

capacitor and the resistive load. The circuit shown in Fig. 8.30 is a half-wave rectifier with a parallel connection of a resistance R_c element first of the diode D1. The resistance is placed in that position because it reduces the charging time of the output capacitor. The optimal value of R_c is what makes it possible to reach the final voltage on the capacitor C_2 without load in a minimum time. The circuit shown in Fig. 8.31 is a full-wave rectifier that works as a voltage doubler. At the resonance frequency, the piezoelectric element capacitance dominates its internal resistance, and therefore the rectifier operates as a voltage doubler with a fixed capacitor piezoelectric element. In the circuit shown in Fig. 8.32, the diodes are replaced by transistors which operated in a synchronous mode in contrast with the Fig. 8.30. In this solution, when no initial energy is available in the system, can be started and then supplied power to the operational amplifiers to start synchronous operation. The comparator U2 provides a high signal that turns on the transistor M2 when the input voltage has a negative value. The efficiency obtained by using a load of 80 kΩ in the circuits shown in Figs. 8.30, 8.31, and 8.32, is 34 %, 57 %, and 92 %, respectively.

8.10 Electrical Storage Buffer

After the recovery and conversion of the energy, is needed to accumulate it through an appropriate (storage) system. Typically, for a variety of electro-mechanical applications, there are three main storage technologies:

- Fuel cells, mainly used for electric vehicles. A fuel cells is an electrochemical device that allows to obtain electricity directly from certain substances, such as oxygen or hydrogen, which takes place without any thermal combustion process. The efficiency can be very high with a considerable mechanical stability; some processes such as catalysis, however, pose practical limits to their efficiency (Fig. 8.33).

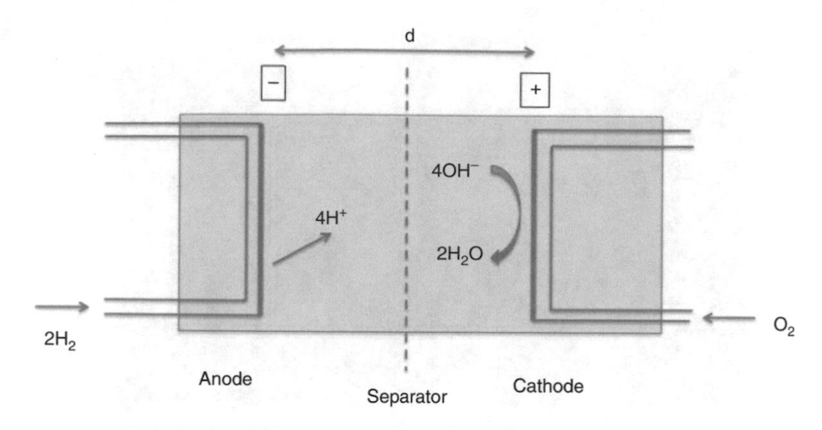

Fig. 8.33 Fuel cell

- Electrochemical batteries, commonly used for electric vehicles, plug-in, and as accumulators for stand-alone photovoltaic systems. The principle of operation of an electrochemical battery is not much different than that of a fuel cell, since the structure consists of two electrodes, each of which is affected by a redox reaction, separated by a membrane permeable of ions. A difference of a fuel cell, the electrochemical batteries are characterized by the presence within the cell of reagents and products. There are various types of electrochemical batteries, including lead-acid, the Nickel-Cadmium, and Lithium ion battery.
- Supercapacitors, or special capacitors that have the characteristic of storing a quantity of exceptionally high electrical charge respect to traditional capacitors; in fact, while the capacitor has values of capacity at the order of mF, supercapacitors can reach up to over 5000 Farad. Therefore, they constitute the storage devices characterized by high specific power and energy by far higher than traditional capacitors.

The electrochemical capacitor observed from Fig. 8.34 is characterized by a very similar construction to that of a battery; it has substantially two electrodes and an ion permeable separator, placed between the electrodes which contains the electrolyte.

A storage system in supercapacitors must be suitably dimensioned in such a way that this has long life and high efficiency; in this regard of particular importance, the parameter known as Dept of Discharge (DOD) expresses how much energy has been withdrawn from the supercapacitor as a percentage of total capacity. Typically, the DOD of a supercapacitor should not be more than 20 % to obtain an optimal operation. The supercapacitors may be classified according to the material used for the construction of the electrodes (carbon, metal oxides, or polymers) and according to the type of electrolyte (organic, aqueous, or solid); the choice of the type of electrolyte will condition then all the dimension of the supercapacitor.

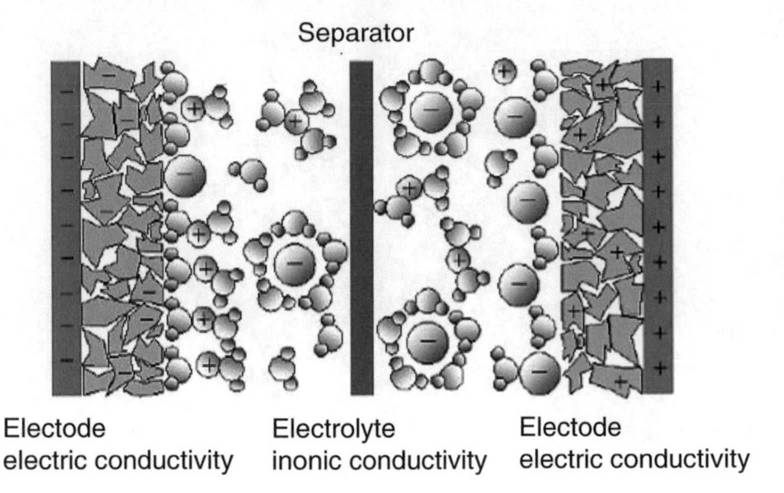

Fig. 8.34 Internal structure of a supercapacitor

8.10.1 *Supercapacitors*

Recent studies in the field of super-capacitors (SC), also known as ultracapacitors or electrochemical capacitors, have developed most viable alternative for the energy storage technologies. These devices in fact possess a greater number of charge-discharge cycles than conventional batteries and can operate in a wide temperature range. Conversely, disadvantages of these devices are the low energy density and high self-discharge. In general, to maximize energy storage efficiency ($\eta_{E\text{storage}}$), the total leakage power is kept low, charging the supercapacitors alternately or according to their voltage series. To maximize energy efficiency on the load ($\eta_{E\text{driving}}$), the residual energy is minimized by changing to a supercapacitor series configuration so as to increase the voltage above the minimum of the dc-dc converter input voltage. Experimental results show that for the charge is preferable the series type when the input power is high due to low losses, while the individual type is preferred when the input power is low. The series type is effective in reducing the residual energy while the individual type can minimize the losses. The energy stored is proportional to V_2, which means that two voltage conversions are usually required: a transducer from energy to charge the super capacitor, the other from the SC to supply the load. The residual energy below the minimum voltage of the dc-dc converter or charge pump normally is not usable and grows linearly with the value of the capacitance and quadratically with the voltage value. Another problem is the leakage current that grows exponentially in the vicinity of the nominal voltage (2.5–2.7 V); typically the supercapacitors have a voltage range from 0 to 2.7 V.

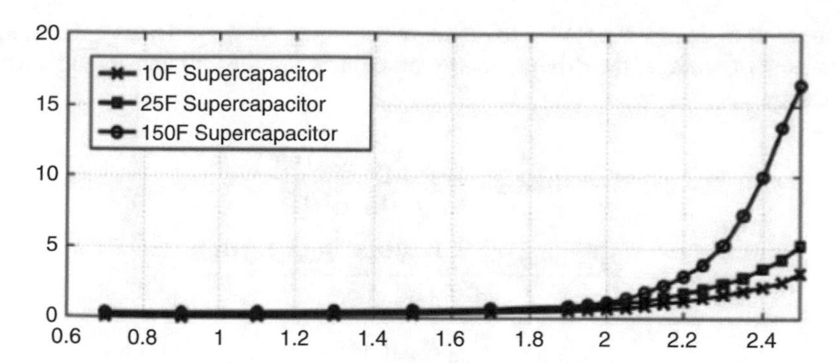

Fig. 8.35 Leakage power loss (mW) in the SC as function of the supercapacitor voltage (V)

The leakage current and power are given by the following expressions:

$$I_l(t) = C\frac{V(t) - V(t + \Delta t)}{\Delta t} \tag{8.5}$$

$$P_l(t) = I_l(t)\Delta V(t) = C\frac{\Delta V^2(t)}{\Delta t} \tag{8.6}$$

From these equations it is possible to derive the energy loss due to leakage:

$$E_l(t) = \int_0^t P_l(t)dt \tag{8.7}$$

The Fig. 8.35 shows how the power of the leakage increases with voltage across of the supercapacitor, in particular when it is closer to the nominal voltage.

In addition, the following variables can be defined:

- P_{charge}: power supplied by the transducer (at time t), assumed maximum and converted to the appropriate voltage, which can be collected by the SC or less. If it is not accepted, then it is dissipated as heat.
- P_{overhead}: sum of the conduction losses and switching losses.
- E_{held}: energy possessed by the supercapacitor (at time t) less leakage and switching losses.
- E_{residual}: power unusable in SC with voltage below the minimum threshold of the dc-dc converter. item P_{LOAD}: power supplied to the load from the SC to the time t.

In terms of power is possible to define the charging power efficiency (η_c) and the driving power efficiency (η_d) as follows:

$$\eta_c = \frac{P_{\text{accepted}}}{P_{\text{changed}}} \tag{8.8}$$

$$\eta_d = \frac{P_{\text{load}}}{P_{\text{discharge}}} \tag{8.9}$$

In terms of energy is possible to define the charging energy efficiency, the energy efficiency of storage, the driving energy efficiency, and overall end-to-end energy efficiency:

$$\eta_{E\text{charging}} = \frac{\int_0^t P_{\text{accepted}}(t)dt}{\int_0^t P_{\text{charge}}(t)dt} \tag{8.10}$$

$$\eta_{E\text{storage}} = \frac{\int_0^t (P_{\text{accepted}}(t) - P_{\text{leak}}(t) - P_{\text{overhead}}(t))dt}{\int_0^t P_{\text{accepted}}(t)dt} \tag{8.11}$$

$$\eta_{E\text{driving}} = \frac{\int_0^t P_{\text{load}}(t)dt}{\int_0^t P_{\text{discharge}}(t)dt + \int_0^t P_{\text{overhead}}(t)dt + E_{\text{residual}}(t)} \tag{8.12}$$

There are three types of Energy Storage Elements (ESE) for energy harvesting (EH) systems:

- Single Supercapacitor (SS): A Single Small Supercapacitor (SSS, about 1 F) can charge quickly with a charging time long.
- Reservoir Supercapacitor Array (RSA): an array of SSS which allows to reduce the losses of leakage.
- Dynamic Reconfigurable Supercapacitors: This type allows to reduce the leakage and E_{residual} and improves the efficiency of the output stage.

References

1. Park, J., & Mackqy, S. (2003). *Practical data acquisition for instrumentation and system control*. Amsterdam: Elsevier.
2. Lacanette, K. (2003). National temperature sensors handbook. Handbook Ann. Mat. National semiconcductor.
3. National Instruments (1996). Data acquisition fundamentals, application note 007 (1996).
4. National Instruments (1996). Signal conditioning fundamentals for PC-based data acquisition systems, Handbook National Instruments
5. Taylor, J. (1986). Computer-based data acquisition system. Instrument Society of America, USA.
6. Di Paolo Emilio, M. (2013). *Data acquisition system, from fundamentals to applied design*. New York: Springer.
7. Roundy, S., Wright, P., & Pister, K. (2002). *Micro-electrostatic vibration-to- electricity converters*. In Proceedings of ASME International Mechanical Engineering Congress and Exposition IMECE2002 (Vol. 220, pp. 17–22).
8. Stordeur, M., & Stark, I. (1997). *Low power thermoelectric generator: Self- sufficient energy supply for micro systems*. In Proceedings of the 16th International Conference on Thermoelectrics (pp. 575–577).
9. Shenck, N., & Paradiso, J. (2001). Energy scavenging with shoe-mounted piezoelectrics. *IEEE Micro, 21*(3), 30–42.
10. Roundy, S. (2003). *Energy scavenging for wireless sensor nodes with a focus on vibration to electricity conversion*. PhD thesis, University of California.

11. Tsutsumino, T., Suzuki, Y., Kasagi, N., Kashiwagi, K., & Morizawa, Y. (2006). Micro seismic electret generator for energy harvesting In *Technical Digest PowerMEMS 2006*, Berkeley, USA, November 2006 (pp. 133–136).
12. Sterken T., Altena G., Fiorini P., & Puers, R. (2007) Characterisation of an electrostatic vibration harvester, DTIP of MEMS and MOEMS, Stresa, Italy, April 2007.
13. Sterken, T., Baert K., Puers, R., & Borghs, S. (2002). Power extraction from ambient vibration. In *Proceedings of the SeSens*, Workshop on Semiconductor Sensors, Veldhoven, Netherlands, November 2002 (pp. 680–683).
14. Szarka, G., Stark, B., & Burrow, S. (2012). Review of power management for energy harvesting systems. *IEEE Transactions on Power Electronics, 27*(2), 803–815. ISSN:0885-8993.
15. Cammarano, A., Burrow, S. G., Barton, D. A. W., Carrella, A., & Clare, L. R. (2010). Tuning a resonant energy harvester using a generalized electrical load. *Journal of Smart Materials and Structure, 19*, 055003.
16. Guyomar, D., Badel, A., Lefeuvre, E., & Richard, C. (2005). Toward energy harvesting using active materials and conversion improvement by nonlinear processing. *IEEE Transactions on Ultrasonics, Ferroelectrics, and Frequency Control, 52*, 584–595.
17. Mitcheson, P. D., Stoianov, I., & Yeatman, E. M. (2012). Power-extraction circuits for piezoelectric energy harvesters in miniature and low-power applications. *IEEE Transactions on Power Electronics, 27*, 4514–4529.
18. Szarka, G. D., Burrow, S. G., & Stark, B. H. (2012). Ultra-low power, fully-autonomous boost rectifier for electro-magnetic energy harvesters. *IEEE Transactions on Power Electronics, 28*(7), 3353–3362. doi:10.1109/TPEL.2012.2219594.
19. Maurath, D., Becker, P. F., Spreeman, D., Manoli, Y. (2012). Efficient energy harvesting with electromagnetic energy transducers using active low-voltage. *IEEE Journal of Solid-State Circuits, 47*(6), 1369–1380.
20. Beeby, S. P., Tudor, M. J., & White, N. M. (2006). Energy harvesting vibration sources for microsystems applications. *Measurement Science and Technology, 17*, R175–R195.
21. Khaligh, A., Zeng, P., & Zheng, C. (2010). Kinetic energy harvesting using piezoelectric and electromagnetic technologies - state of the art. *IEEE Transactions on Industrial Electronics, 57*(3), 850–860.
22. Paulo, J., & Gaspar, P. D. (2010). Review and future trend of energy harvesting methods for portable medical devices. In *Proceedings of the World Congress on Engineering* (Vol. 2).
23. Zhu, D., Tudor, M. J., & Beeby, S. P. (2010). Strategies for increasing the operating frequency range of vibration energy harvesters: A review. *Measurement Science and Technology, 21*, 022001-1–022001-29.
24. Cepnik, C., Lausecker, R., & Wallrabe, U. (2013). Review on electrodynamic energy harvesters - a classification approach *Micromachines, 4*(2), 168–196. http://www.mdpi.com/2072-666X/4/2/168. Accessed 20 January 2015.
25. Ulaby, F. T., Michielssen, E., & Ravaioli, U. (2010). *Fundamentals of applied electromagnetics* (6th ed.). Upper Saddle River: Prentice Hall, Oxford.
26. Roundy, S., Wright, P. K., & Rabaey, J. M. (2003). A study of low level vibrations as a power source for wireless sensor nodes. *Computer Communications, 26*(11), 1131–1144.
27. Sazonov, E., Li, H., Curry, D., & Pillay, P. (2009). Self-powered sensors for monitoring of highway bridges. *IEEE Sensors Journal, 9*, 1422–1429.
28. Toh, T. T., Mitcheson, P. D., Holmes, A. S., & Yeatman, E. M. (2008) A continuously rotating energy harvester with maximum power point tracking. *Journal of Micromechanics and Microengineering, 18*, 104008-1-7.
29. Howey, D. A., Bansal, A., & Holmes, A. S. (2011). Design and performance of a centimetre-scale shrouded wind turbine for energy harvesting. *Smart Materials and Structures, 20*, 085021.
30. Razavi, B. (2002). *Design of analog CMOS integrated circuits*. New York: McGraw-Hill.
31. Razavi, B. (2008). *Fundamentals of microelectronics*. London: Wiley

32. Sedra, A. S., & Smith, K. C. (2013). *Microelectronic circuits*. Oxford: Oxford University Press.
33. Razavi, B. (2002). *Design of integrated circuits for optical communications*. New York: McGraw-Hill.
34. Hurst, P. J. (2001). *Analysis and design of analog integrated circuits*. New York: Wiley.
35. Spies, P. (2015). *Handbook of energy harvesting power supplies and applications*. Boca Raton: CRC Press Book, France.

Chapter 9
Low-Power Circuits

9.1 Introduction

Most of the signals applied to the inputs of an electronic system are originated from the devices called sensors which, on the basis of their input–output characteristic, convert non-electrical quantities (for example, a temperature) into corresponding electrical values in analog form (for example, a voltage). Furthermore, most of the signals present at the outputs of an electronic system are used to drive devices called actuators. For this reason, although the processing of the information takes place typically in digital form, all electronic systems require the presence of circuits for processing analog signals, if only to convert these analog signals into digital signals (analog/digital conversion or A/D) and vice versa (D/A conversion). The main operations that can be performed on analog signals are the amplification (which is equivalent to multiplying the signal by a constant), the weighted sum of several signals, the operations of derivation and integration in time, and the filtering in the frequency domain [1–15].

9.2 Review of Microelectronics

The earliest electronic circuits were fairly simple. They were composed of a few tubes, transformers, resistors, capacitors, and wiring. As more was learned by designers, they began to increase both the size and complexity of circuits. Component limitations were soon identified as this technology developed. The transition from vacuum tubes to solid-state devices took place rapidly. As new types of transistors and diodes were created, they were adapted to circuits. The reductions in size, weight, and power use were impressive. Microelectronic technology today includes thin film, thick film, hybrid, and integrated circuits and combinations of these. Such circuits are applied in DIGITAL, SWITCHING, and LINEAR (analog)

M. Di Paolo Emilio, *Microelectronic Circuit Design for Energy Harvesting Systems*, DOI 10.1007/978-3-319-47587-5_9

circuits. Because of the current trend of producing a number of circuits on a single chip, you may look for further increases in the packaging density of electronic circuits. At the same time you may expect a reduction in the size, weight, and number of connections in individual systems. Improvements in reliability and system capability are also to be expected.

9.2.1 Basic of Semiconductor's Physics

Microelectronics is based on physics of structures. The main semiconductor elements used in the applications are silicon, boron, aluminum, etc. Silicon atom has 4 valence electrons and it's important for electrical system. At temperature close to 0 K, silicon crystal behaves as an insulator: its electrons are confined to their, respectively, covalent bonds. At higher temperature, electrons gain thermal energy and can be used as "free charge." Bandgap energy, minimum energy to broke the bond, of the silicon is 1.12 eV where 1 eV $= 1.6 * 10^{-19}$ J. The number of "free electrons" that is possible to have in a semiconductor crystal depends on band gap energy and temperature, it's possible to define density of electrons (number of electrons per unit volume) as the following:

$$n = 5.2 * 10^{15} T^{\frac{3}{2}} * e^{-\frac{E_g}{2KT}} \tag{9.1}$$

with k = boltzmann constant $= 1.38 * 10^{-23}$ J/K. For insulator materials $E_g \simeq$ 2.5 eV, instead, for conductor is about less than 1 eV. As example we can calculate the density of electrons for silicon at $T = 300$ K:

$$n = 5.2 * 10^{15} T^{\frac{3}{2}} * e^{-\frac{1.72*10^{-19}}{2KT}} \sim 10^{10} \text{e/cm}^3 \tag{9.2}$$

The intrinsic semiconductors can't be used for electrical application due to very high resistance but also a small number of "free electrons" with respect to the conductor. In an extrinsic semiconductor the situation is different: there is a p-Si and an n-Si. In the first case majority carriers (holes) are $p = N_A$, semiconductor is doped with a density of N_A (i.e., boron atom) and minority carriers $n = p_i^2/N_A$; in the other case, instead, majority carriers (electrons) $n = N_A$ with semiconductor is doped with a density of N_D (i.e., phosphorus), minority carriers $p = n_i^2/N_D$. The concept of doping extends the flow of current in a semiconductor.

9.2.1.1 Drift

The drift is the movement of electrons charge due to an electric field: $v = \mu E$, μ is "mobility" expressed in $\text{cm}^2/\text{V} \cdot \text{s}$. If consider a piece of n-silicon

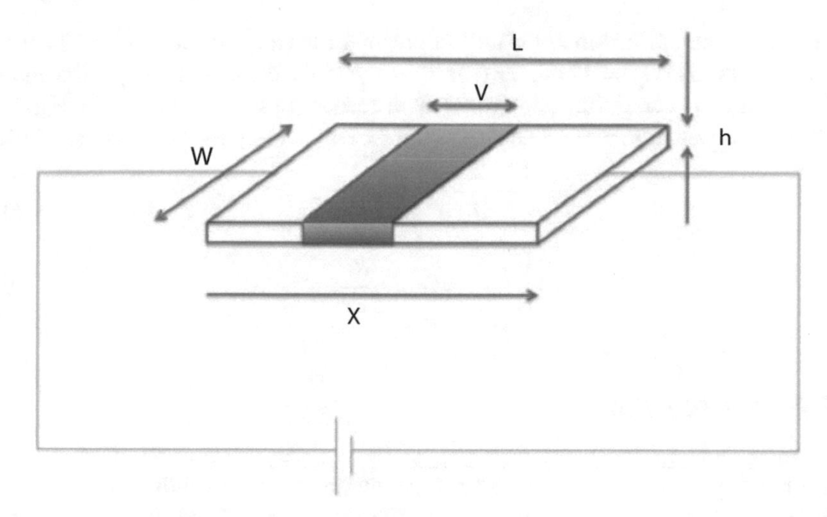

Fig. 9.1 Current flow in terms of charge density

($\mu_n = 1350\,\text{cm}^2/\text{V}\cdot\text{s}$) of $L = 1\,\mu\text{m}$ applied a voltage of $V = 1\,\text{V}$, the velocity of electrons (drift) is about $v = \mu_n E = \mu_n \frac{V}{L} \simeq 10^7\,\text{cm/s}$. For a *p*-silicon $\mu_p = 480\,\text{cm}^2/\text{s}$.

Now, we should calculate the current due to this drift v (Fig. 9.1):

$$I = -v \cdot W \cdot h \cdot n \cdot q \tag{9.3}$$

It can be written in another way:

$$J_n = \mu_n \cdot E \cdot n \cdot q \tag{9.4}$$

where J is the current density, $W \cdot h$ is the cross section, and $n \cdot q$ is the charge density in column of the semiconductor. In presence of both electrons and holes:

$$J = J_n + J_p = q(\mu_n n + \mu_p p)E \tag{9.5}$$

9.2.1.2 Diffusion

The mechanism of the diffusion is the movement of the charge from a zone of high concentration to zone of low concentration (current). The mathematical relation that explains this phenomena is the following:

$$I = A \cdot q \cdot D \cdot n \frac{dn}{dx} \tag{9.6}$$

where D_n is the "diffusion constant" in cm^2/s; in intrinsic silicon $D_n = 34\,cm^2/s$ (for electrons) and $D_p = 12\,cm^2/s$ (for holes). A is the cross section, q is the charge, and $\frac{dn}{dx}$ is the concentration (electrons) with respect to the direction "x" (Fig. 2.1). From the last equation we can calculate the current density for electrons and holes:

$$J_n = qD_n\frac{dn}{dx} \tag{9.7}$$

$$J_p = -qD_p\frac{dp}{dx} \tag{9.8}$$

9.2.2 PN Junction

From the doping we have obtained p-type and n-type semiconductors. A electric field or a concentration gradient leads to the movement of the charges (electrons and holes). We suppose to dope two adjacent pieces of semiconductors (Fig. 9.2); in this configuration there will be a flow of electrons from n to p side and a flow of holes in opposite direction (Fig. 9.3).

At the end of this process of "equilibrium," an electric field (depletion region) will emerge as indicated in Fig. 9.3. The junction reaches equilibrium once the

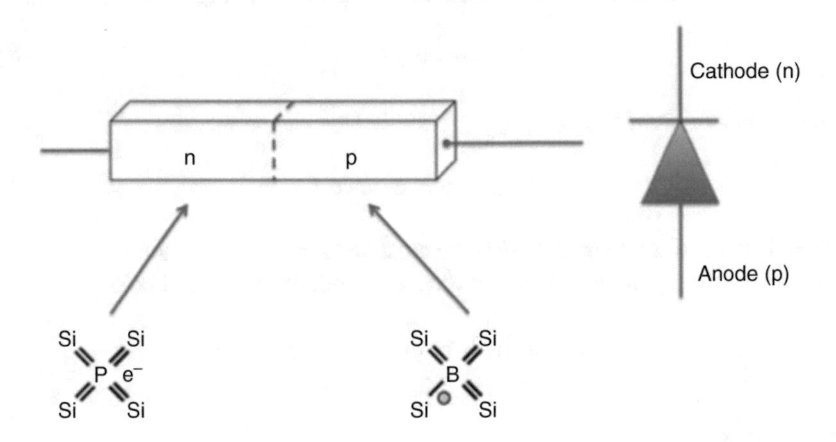

Fig. 9.2 Current flow in terms of charge density

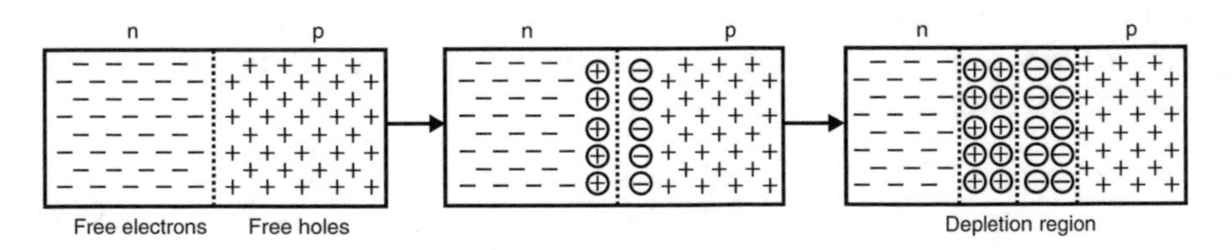

Fig. 9.3 Evolution of charge concentration in a PN junction

electric field is strong enough to completely stop the diffusion current [16–25]. The existence of electric field due to depletion region suggests that PN junction has a built-in potential defined in the following equation:

$$V(x_2) - V(x_1) = -\frac{D_p}{\mu_p} \ln\left(\frac{p_p}{p_n}\right) \tag{9.9}$$

Considering the Einstein's equation, $\frac{D}{\mu} = \frac{\mathrm{KT}}{q}$:

$$|V_0| = \frac{\mathrm{KT}}{q} \ln\left(\frac{p_p}{p_n}\right) \tag{9.10}$$

where p_n and p_p are the concentrations at x_1 and x_2, respectively. The PN junction so realized is called diode. Having analyzed the PN junction in equilibrium, let us observe how it works applying an external voltage (Fig. 9.4).

9.2.2.1 Reverse Bias

The external voltage rises the electric field of depletion region prohibiting the flow of current. The device works as capacitor:

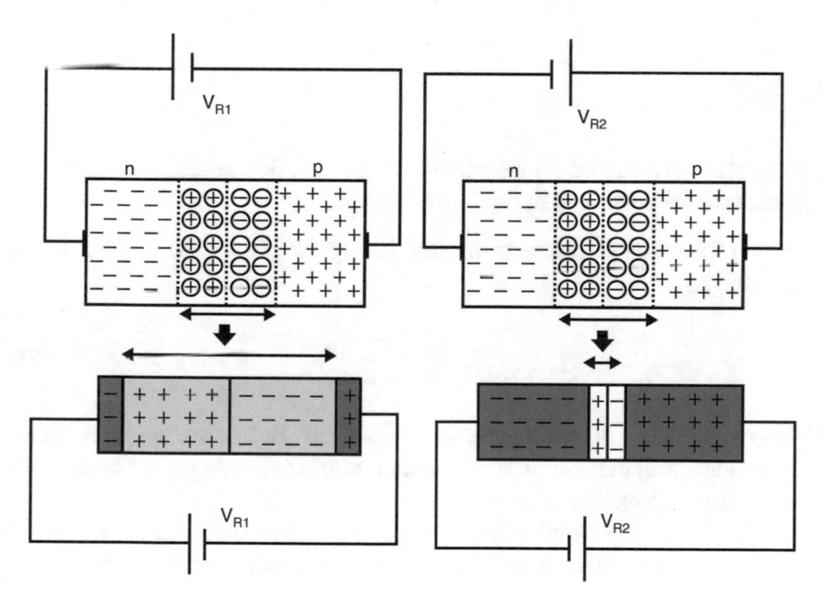

Fig. 9.4 PN junction with external voltage

$$C_j = \frac{C_{j0}}{\sqrt{1 - \frac{V_R}{V_0}}}; C_{j0} = \sqrt{\frac{q\epsilon_{s_i}q}{2} \frac{N_A N_D}{N_A + N_D} \frac{1}{V_0}} \tag{9.11}$$

ϵ_{s_i} is the dielectric constant of silicon.

9.2.2.2 Forward Bias

The external voltage decreases the electric field of the depletion area allowing greater diffusion current (Fig. 9.4):

$$I_D = I_s \left(\exp\left(\frac{V_R}{V_T}\right) - 1 \right) \tag{9.12}$$

with V_T, thermal voltage, and I_s:

$$I_s = \text{Aqn}_i^2 \left(\frac{D_n}{N_A L_n} + \frac{D_p}{N_D L_p} \right) \tag{9.13}$$

is the "reverse saturation current" and L_n and L_p are the electrons and holes "diffusion length" (i.e., $L_n = 20\,\mu\text{m}$, $L_p = 30\,\mu\text{m}$)

9.2.3 *Diode*

The diode is a two terminal device with I-V characteristic indicated in Fig. 9.5. Some application can be described in the following text:

- Wave rectifier: the circuit is visualized in Fig. 9.6: the ripple amplitude can be calculated by:

$$V_r \sim \frac{V_p - V_{D,\text{on}}}{R_L C_1 f_{\text{in}}} \tag{9.14}$$

- Voltage regulation: due to significant variation of the line voltage, it is necessary to be produced in the output of the diode a stabilized voltage. A possible outline is visualized in Fig. 9.7.
- Limiting circuit: the circuit passes the input to the output, $V_{\text{out}} = V_{\text{in}}$ and when the input exceeds a "threshold" the output remains constant (Fig. 9.8).

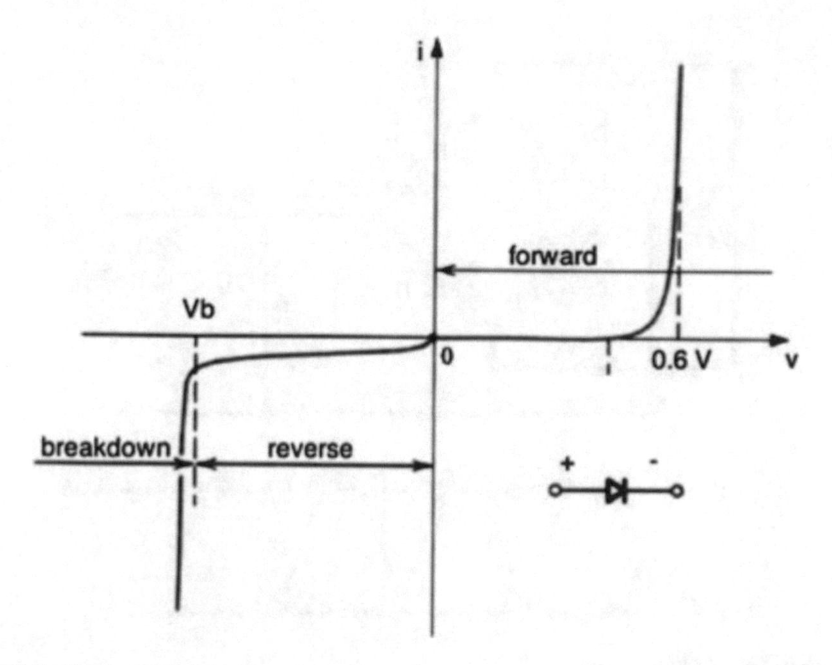

Fig. 9.5 Characteristic I-V of the diode

9.2.4 Bipolar Transistor: Emitter Follower

The union of two junctions *p–n* (i.e., two diodes together) form the bipolar transistor junction (BJT). Bipolar because the current is sustained by electrons and holes (such as the diode). Compared to the diode, the BJT (three terminals) can be used as a signal amplifier. Although the MOS technology (see next paragraph) is more widespread, the technology bipolar remains significant (or predominant, in certain cases) in several applications:

- Electronics vehicle
- Systems wireless
- Digital circuits ECL
- Draft discrete circuits.

The emitter follower circuit is particularly useful for applications where high input impedance is required. It is typically used as a buffer in a wide variety of areas. Emitter follower is a common collector transistor configuration. It can easily be designed by circuit RC. The emitter follower is also known as a voltage follower, or a negative current feedback circuit, with high input impedance and low output impedance. The outline of the emitter follower is shown in Fig. 9.9 and the corresponding equivalent small signal circuit in Fig. 9.10. We can calculate the effective resistance seen from terminal B by:

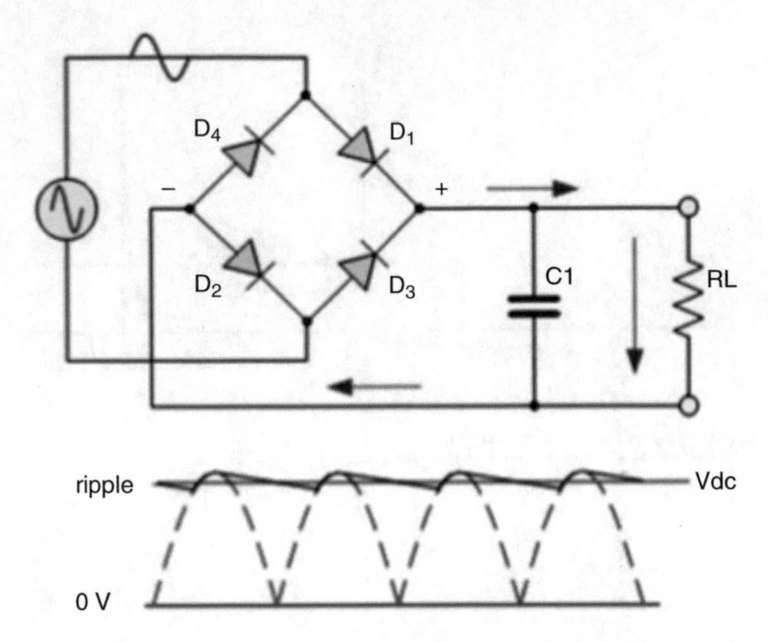

Fig. 9.6 Wave rectifier

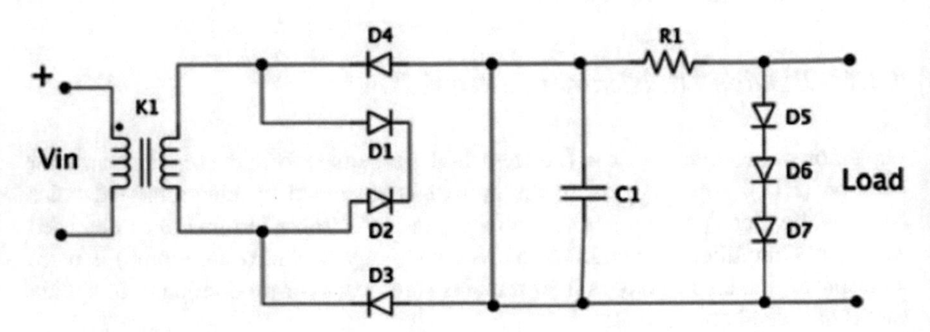

Fig. 9.7 Block diagram of voltage regulator with diodes

$$R_{ib} = r_\pi + (\beta + 1)R_E//R_L \simeq (\beta + 1)R_E//R_L \tag{9.15}$$

This value is relatively high. In general, $R_E//R_L$ is around kΩ and $\beta \sim 100$, so the resistance seen from terminal B is in the hundreds of kΩ. It is evident from Fig. 9.10 that input resistance depends on load resistance. Let us named the input resistance as R_{in}:

$$R_{\text{in}} = R_B//R_{ib} \tag{9.16}$$

The effective resistance is the parallel between R_{ib} and R_b. A common collector configuration can be used as an amplifier in such a circuit, where a large input

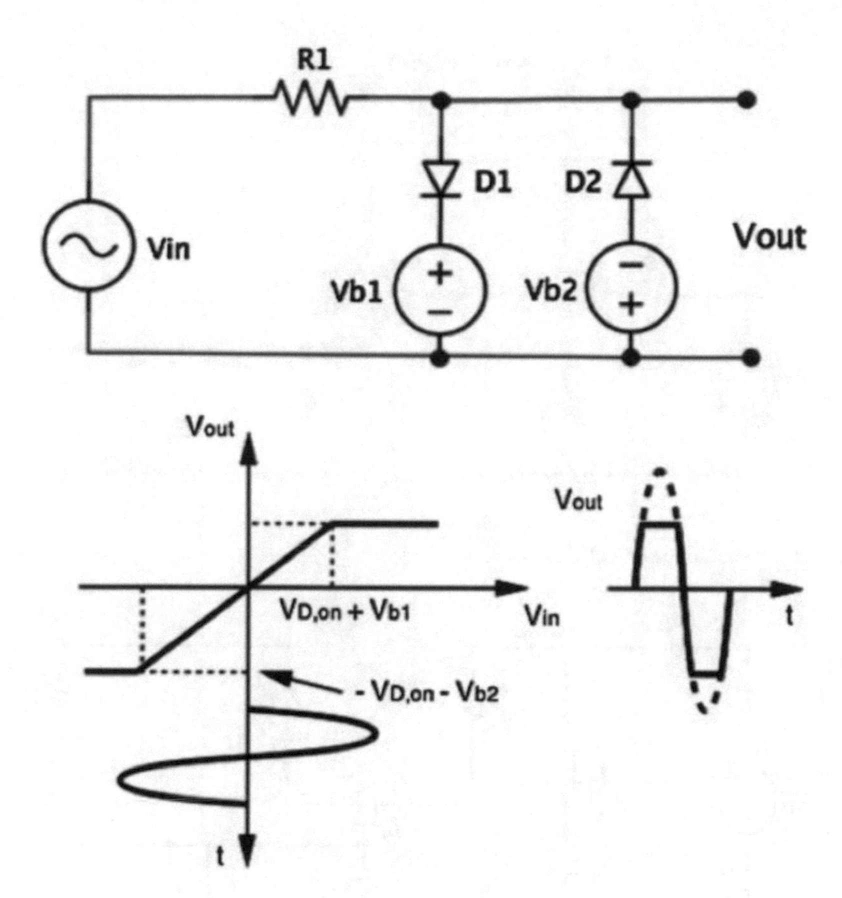

Fig. 9.8 Limiting circuit

resistance is needed; a good application can be a pre-amplifier circuit. From the voltage gain equation ($A = V_0/V_i$) follows:

$$A = \frac{V_0}{V_s} = \frac{R_E//R_L}{r_e + (R_E//R_L)} * \frac{R_{\text{in}}}{R_s + R_{\text{in}}} \tag{9.17}$$

It becomes evident that as long as $r_e \ll (R_E//R_L)$ and $R_s \ll R_{\text{in}}$ the gain approaches unity.

The emitter follower can be designed using the main steps described below:

- Choose a transistor: the transistor should be selected according to the system requirements.
- Select emitter resistor: selecting a working point (for example, select an emitter voltage of about half the supply voltage).
- Determine the base current: base current is the collector current divided by β (or h_{fe}).

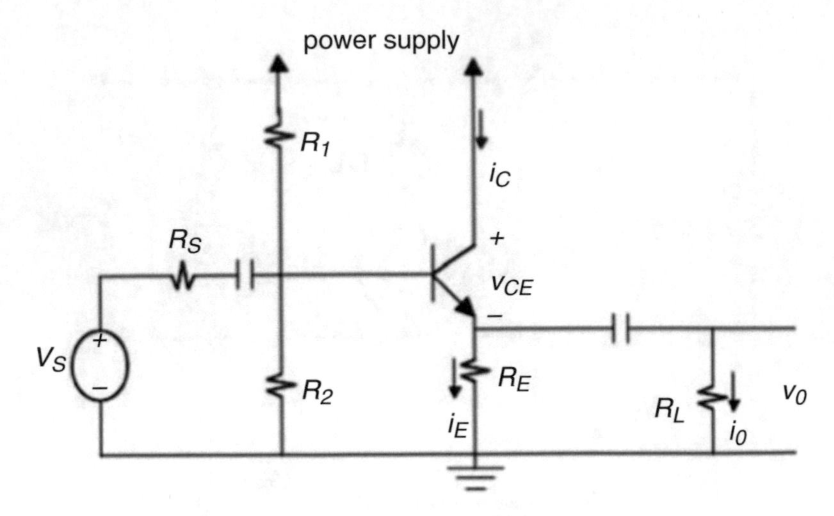

Fig. 9.9 Emitter follower (outline)

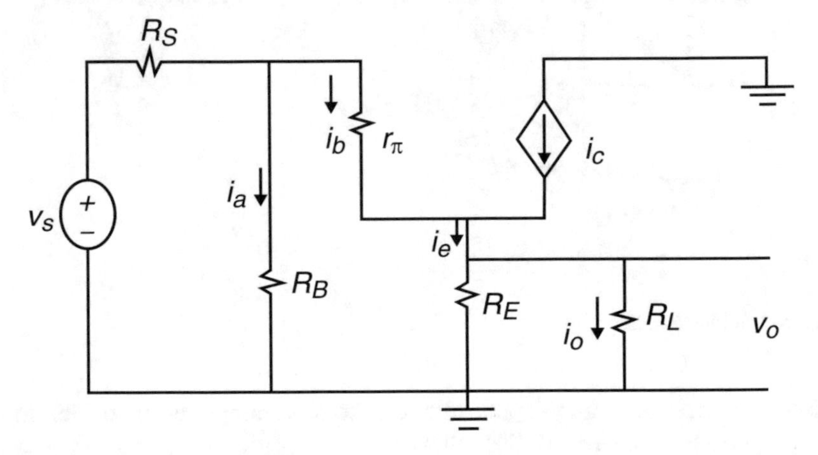

Fig. 9.10 Emitter follower (small signal equivalent circuit)

- Determine the base resistor values: select the value of the resistor(s) to provide the voltage required at the base.
- Determine the value of the input and output capacitor: The value of the input/output capacitor should equal to the resistance of the input/output circuit at the lowest frequency of operation.

When using the emitter follower circuit, there are two main practical points to note:

- The collector may need decoupling: in some cases the emitter follower may oscillate, in particular if long leads are present. One of the easier ways to prevent

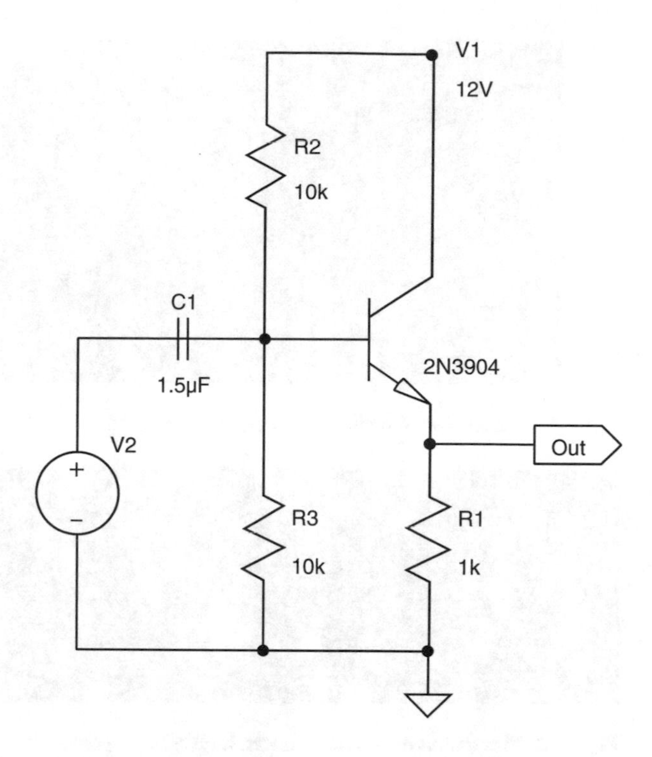

Fig. 9.11 Emitter follower (example with 2N3904)

this is to decouple the collector to ground with very short connections, or by placing a small resistance between the collector and the power supply line.

- The input capacitance affects the RF: the base emitter capacitance may reduce the high impedance of the input circuit if the signal is above 100 kHz.

One of the designs of an emitter follower is shown in Fig. 9.11, with a *P*-spice simulation in Figs. 9.12 and 9.13. This configuration can be typically used in AC coupling applications. It's important that the frequency response and bias voltage on the base are planned for. In the circuit mentioned here, there's an added voltage divider, consisting of R_2 and R_3, and an AC coupling capacitor, C_1. Their component values will need to be calculated. An important aspect of the emitter follower is its relative immunity to temperature instability. When the current in the collector increases (changing of β), the voltage across the emitter resistor also increases. This acts as a negative feedback, since it reduces the voltage difference between base and emitter (V_{BE}), which drives the transistor to conduct more current, as such thermal stability is maintained [26–30].

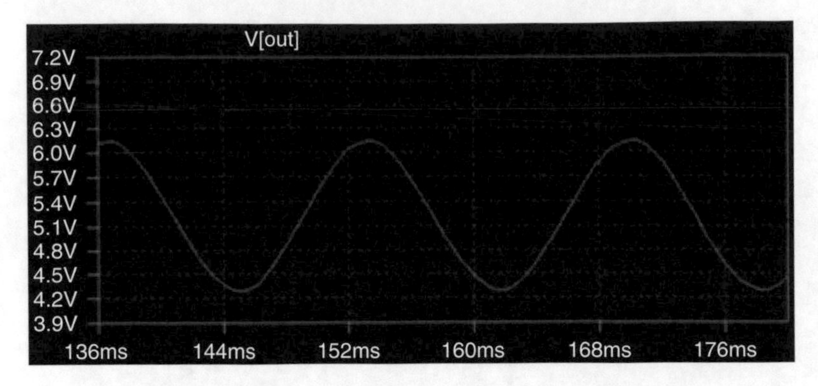

Fig. 9.12 Simulation of emitter follower, output voltage

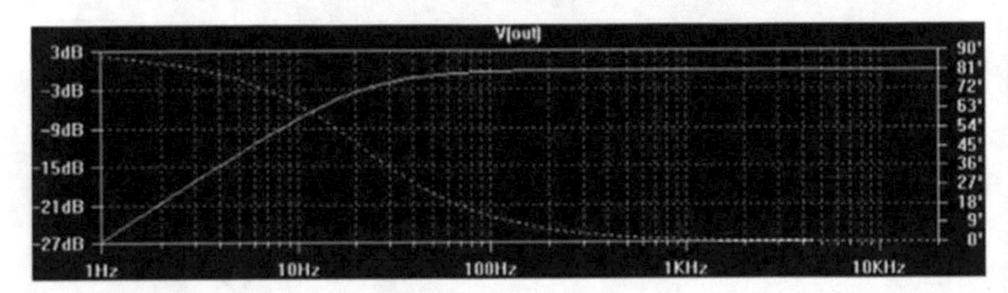

Fig. 9.13 Simulation of emitter follower, frequency response

9.2.5 MOS Transistor

Today the microelectronics is dominated by MOSFET devices. The physical structure of the MOSFET is described in Fig. 9.14: on a substrate of monocrystalline *p*-type, two junctions are made of type $n+$ which are connected by two terminals called drain and source (*D* and *S* in Fig. 9.14). In the area between the drain and source is made to grow a layer of silicon dioxide as excellent insulator thick less than 0.01 m.

The gate terminal is isolated from the Si substrate by a thin layer of SiO_2, and therefore the gate current DC is null. The body terminal is generally connected to that source. Drain and source are symmetrical.

Applying a sufficiently positive voltage between the gate and source, the electrons are attracted towards the $Si - SiO_2$ interface under the gate forming a conductive channel between source and drain. In these conditions, if it is applied a $V_{ds} > 0$, a current may flow between the drain and source. The current between drain and source is controlled by the voltage between gate and source, which controls the formation of the channel. The electrical characteristics of the MOSFET depend from L (gate length) and W (gate width), as well as technological parameters such as oxide thickness and doping of body. Typical values of L and W are: $L = 0.1 - 2\,\mu\text{m}$, $W = 0.5 - 500\,\mu\text{m}$. The range of gate oxide thickness is 3–50 nm.

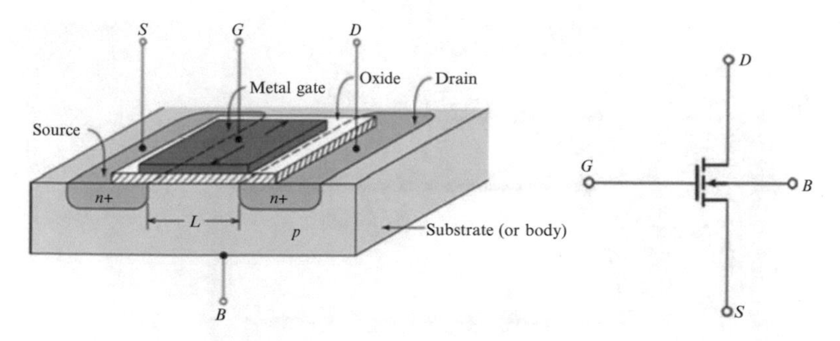

Fig. 9.14 Inside outline (general) and symbol of Mosfet

There are four types of MOS transistors: two to channel n and two to channel p. The MOSFET n-channel (nMOS) is formed on a p-type substrate:

- nMOS enrichment (enhancement) or normally off
- nMOS depletion (depletion) or normally on.

The MOSFET p-channel (pMOS) is formed on a substrate of n-type:

- pMOS enrichment (enhancement) or normally off
- pMOS depletion (depletion) or normally on.

In most MOSFET applications, an input signal is the gate (G) voltage V_g and the output is the drain (D) current I_d. The ability of MOSFET to amplify the signal is given by the output/input ratio: the transconductance, $g_m = dI/dV_{\text{gs}}$.

In Fig. 9.15 reports the characteristic curve $I_d - V_{\text{ds}}$; there are three regions of work:

- Cutoff: in this case its necessary induce the channel, $V_{\text{gs}} \leq V_t$ (V_t is the threshold voltage) for nMOS.
- Triode: the channel must be induce and also keep V_{ds} small enough so the channel is continuous (not pinched off): $V_{\text{ds}} \leq V_{\text{gs}} - V_t$.
- Saturation: in this mode need to induce the channel, $V_{\text{ds}} \geq V_{\text{gs}} - V_t$ then ensure that the channel is pinched off at the drain end. In Fig. 9.16 is visualized the plot of I_d versus V_{gs} for an enhancement type nMOS device in saturation.

In the saturation mode, this device is an ideal current source (Fig. 9.17). In reality, there is to consider a finite output resistance (r_0); the outline of Fig. 9.18 can be described as in Fig. 9.17.

While the transconductance g_m gives the variations of the drain current due to variations of the voltage V_{gs}, there is another fundamental parameter of the Mosfet that takes the name of the output conductance, which is the variation of the drain current I_{ds} due to variations of the voltage V_{ds}.

The output conductance assumes a much greater importance, and the reason is the effect of the so-called channel length modulation due to V_{ds}: it is the effect for which the effective length of the channel decreases with increasing V_{ds} and in

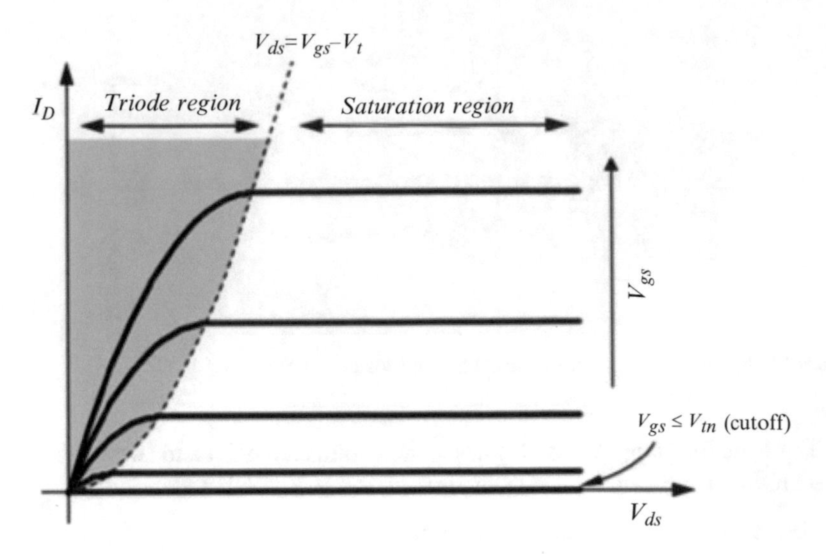

Fig. 9.15 Characteristic curve $I_d - V_{\mathrm{ds}}$

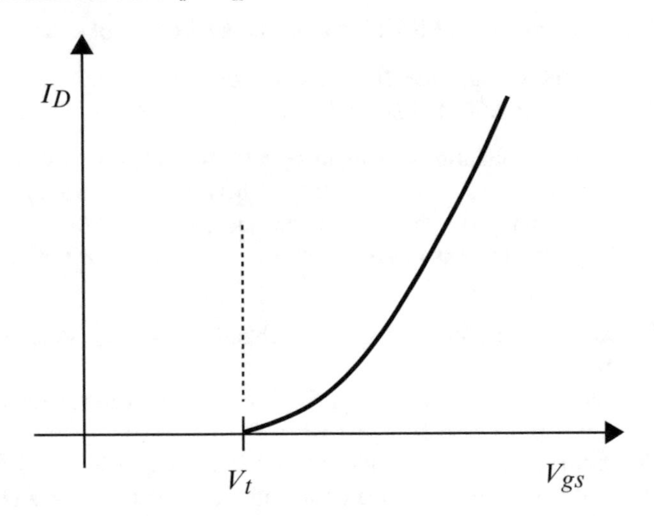

Fig. 9.16 Saturation, $I_d - V_{\mathrm{gs}}$

the saturation zone, the current increases with the V_{ds}. From an analytical point of view the effect of the modulation of the channel length can be written as follows: $I_{\mathrm{ds}} = k(V_{\mathrm{gs}} - V_t) * 2 * (1 + V_{\mathrm{ds}}\lambda)$. It is clear that provides a linear dependence of the I_{ds} vs V_{ds} in according to λ named as parameter of the channel length modulation; k is the transconductance factor proportional to the geometry of the Mosfet.

Example of application with Mosfet can be the common source amplifier visualized in Figs. 9.19 and 9.20. The input signal is applied to the gate through the coupling capacitor C_1. The output is on the drain and connected to the load

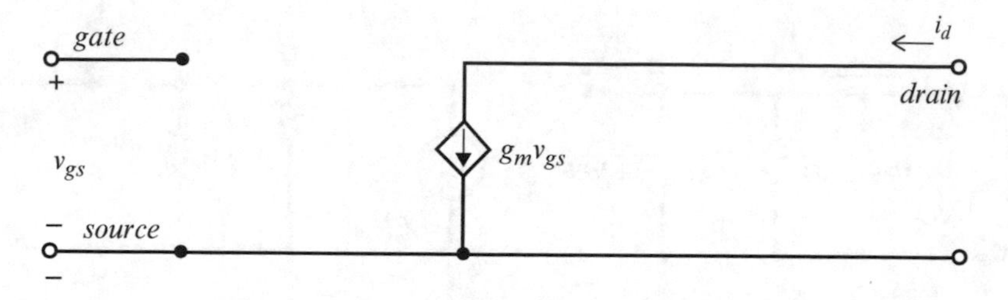

Fig. 9.17 Mosfet model (ideal)

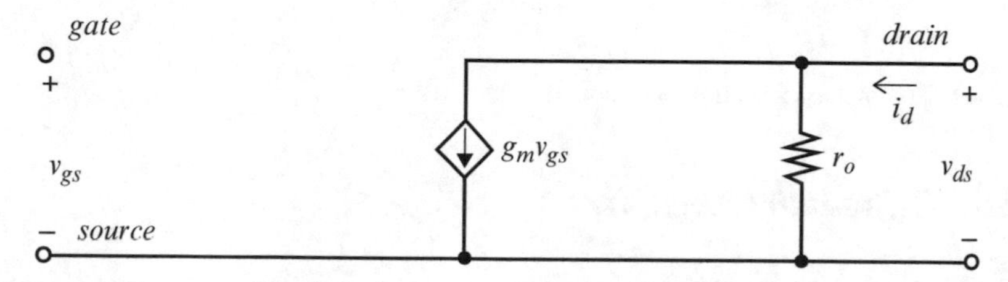

Fig. 9.18 Mosfet model (not ideal)

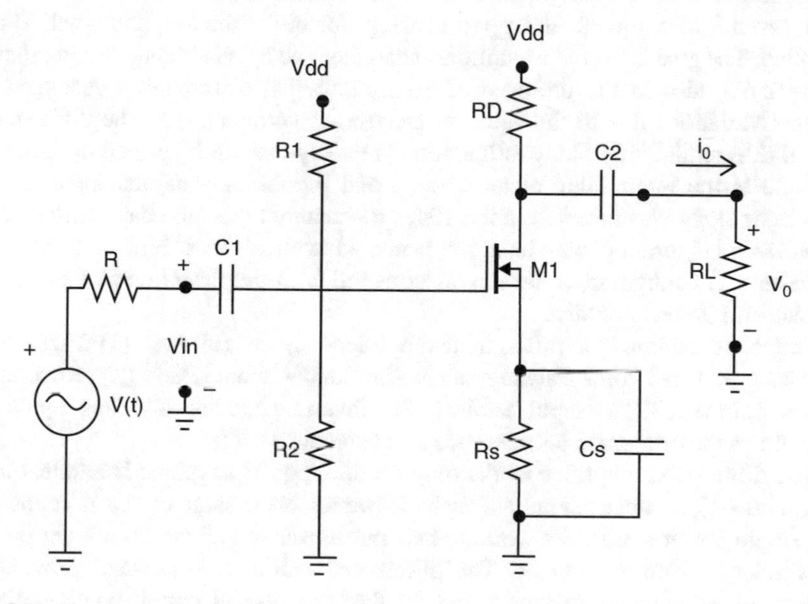

Fig. 9.19 Common source amplifier

by C_2, C_1, C_2, and C_S which are the coupling capacitors and therefore can be considered short circuits at the frequencies of the signal (center-band). The source (S) is therefore to ground for the signals. R_1, R_2, R_D, and R_S form a bias network to four resistors.

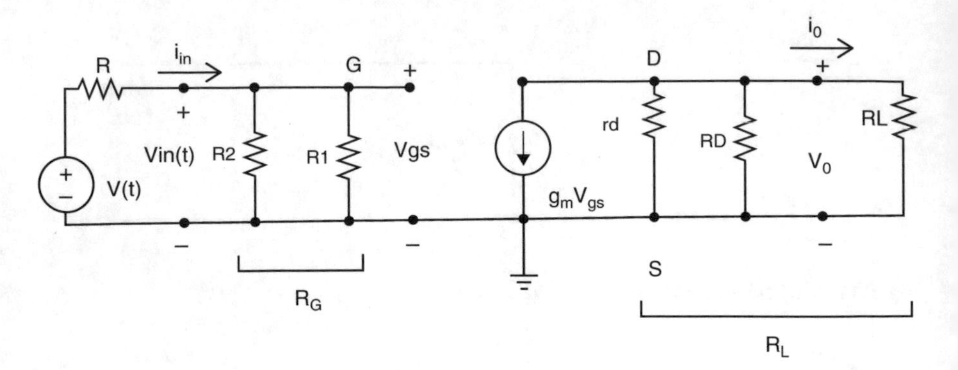

Fig. 9.20 Common source amplifier (model)

9.2.6 *Differential Amplifiers*

The main part of analog-integrated-circuit design is the differential-pair or differential-amplifier configuration. The differential amplifiers were introduced in electronics to eliminate all or part of the problems of the amplifiers with direct coupling. The goal is to create amplifiers characterized by an acceptable signal/noise ratio (SNR), also in the presence of external/internal disturbances generated by thermal variations due to the aging of electronics components. The differential-pair of differential-amplifier configuration is widely used in IC circuit design. One example is the input stage of an op-amp and the emitter-coupled logic (ECL). This technology was invented in the 1940s in vacuum tubes; the basic differential-amplifier configuration was later implemented with discrete bipolar transistors. However, the configuration became most useful with the invention of the modern transistor/MOS technologies.

The main features of differential amplifiers are as follows: (1) High input resistance, so small voltage signals can be amplified without losses. (2) Temperature drift is minimal. (3) Two input terminals, i.e., inverting and non-inverting inputs. (4) It amplifies the difference between two input signals.

The differential amplifier works only on dual power supplies: it requires both $+V_{cc}$ and $-V_{cc}$ voltages simultaneously. However, if the same circuit is connected to a single power supply, its working becomes unstable and we do not get proper amplification from the circuit. The differential amplifier (Fig. 9.21) provides a number of advantages, making it one of the most useful circuit configurations, particularly as input stage for high gain and DC amplifiers. The figure of merit for differential amplifiers is called common-mode rejection ratio (CMRR), defined as the ratio between difference mode gain (Ad) and common-mode gain (Ac). Differential amplifiers is a special purpose amplifier designed to measure differential signals, otherwise known as a subtractor. The CMRR can be calculated with the following equation:

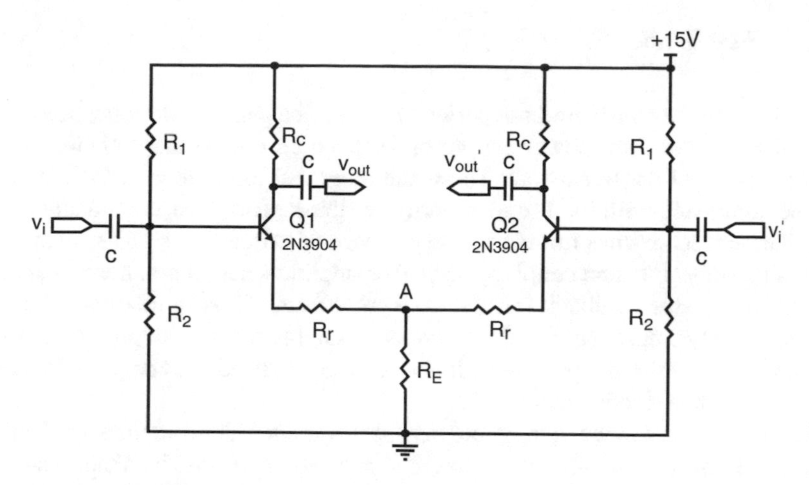

Fig. 9.21 Example of a differential amplifier

$$A_d = \frac{R_c}{2(R_r + r_{\mathrm{tr}})}; A_c = \frac{R_c}{2R_E + R_r + r_{\mathrm{tr}}}; CMRR = \frac{2R_E + R_r + r_{\mathrm{tr}}}{2(R_r + r_{\mathrm{tr}})} \tag{9.18}$$

where r_{tr} is the transresistance, usually indicated also as r_e. The 741 (a common op-amp chip) has a CMRR of 90 dB, which is reasonable in most cases. A value of 70 dB may be adequate for applications insensitive to the effects on amplifier output; some high-end devices may use op-amps with a CMRR of 120 dB or more. The common-mode rejection ratio (CMRR) relates to the ability of the op-amp to reject a common-mode input voltage. This is very important because common-mode signals are frequently encountered in op-amp applications.

Common application of differential amplifiers is in the control of motors or servos, as well as in signal amplification. In discrete electronics, a common arrangement for implementing a differential amplifier is the long-tailed pair, typically found as a differential element in most op-amp integrated circuits. A long-tailed pair can be used as an analog multiplier with a differential voltage as one input and biasing current as another. There are many differential amplifier ICs on the market today. The AD8132 from Analog Devices (ADI) is a low-cost differential (or single-ended) amplifier with one resistor for setting the gain. The AD8132 is a major improvement on op-amps, especially for driving differential input ADCs or signals over long lines. The AD8132 can be used for differential signal processing (gain and filtering) throughout a signal chain, significantly simplifying the conversion between differential- and single-ended components. Linear Technology also provides many differential amplifiers for different applications. For example, its LTC6409 offers a very high speed and low distortion, and is stable in a differential gain of 1.

9.2.7 Feedback

Amplifiers can be considered not perfectly linear. The gain (or amplification) of the amplifier changes with power supply or temperature due to the variations of the working point of the transistors. These and other real limitations of the amplifiers can be minimized with the use of negative feedback. The good functionality of an amplifier and sometimes limited by the presence of extraneous signals, such as the hum of power supply and coupling with other amplifiers neighbors. In some specific cases, the negative feedback can reduce these effects, while in other cases it does not induce any improvement. An important example are the amplifiers and hi-fi audio systems, which are powered with DC voltage obtained by rectifying AC power supply and not perfectly stable.

In Fig. 9.22 is shown a possible feedback outline. The output signal of the amplifier is always smaller than power supply voltages and its shape can be a clear demonstration that the output is not linear. The result of this non-linearity is manifested as a distortion of the output signal.

The use of negative feedback allows to realize amplifiers with better performance. The main effects of feedback loop can be the following:

- greater stability of gain
- less distortion
- greater bandwidth
- reduction of the noise.

In general, the behavior of the individual block of negative feedback can be described by transfer function (Fig. 9.23).

The G block is called forward, the block H is the feedback: the output signal V_{out}(t) is sent back, through H, in input and is controlled by comparator. The difference signal represents the error of the system, it acts as a input signal to the block forward to have in output the corrected signal.

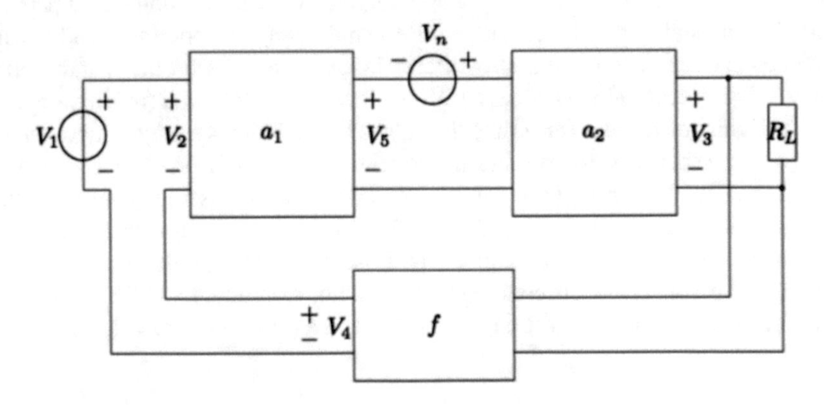

Fig. 9.22 Feedback outline

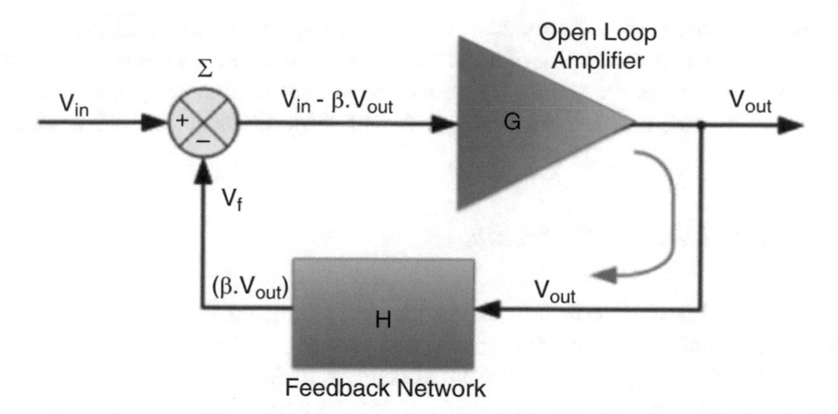

Fig. 9.23 Negative feedback outline (general)

If the system is sufficiently fast, any changes of G block do not affect the output signal. More generally, if the input signal changes in time, the negative feedback system gives the necessary corrections to ensure that the output signal is a faithful repetition (with amplification) of the output.

Otherwise a positive feedback can be used to design oscillators. Positive feedback is responsible for the squealing of microphones when placed too close to the speaker through which their input signals are amplified.

When the loop gain is positive and above 1, there will typically be exponential growth, increasing oscillations or divergences from equilibrium.

9.2.8 Effects of Feedback

A disturbance or noise is an unwanted signal that is superimposed on the input/output of a circuit. Disruptions can occur in various parts of the system, their effect is greater in the input.

Some consideration about the effects of the noise in the loop systems:

- the open loop systems are not able to compensate for the disturbance but if noise is present at the output the effect is smaller because the useful signal which is superimposed has an amplitude greater.
- the closed loop systems are able to limit the effect of the disturbance when this is present in the output but not when it is present at the input.

The effect of disorder can be evaluated through the ratio signal-to-noise (S/N or SNR). The time that elapses between the moment in which it has the effect and the moment in which this effect is taken into account to modify the system is called "delay in the feedback loop." When this delay is high, there may be problems of stability. The Nyquist criterion allows to investigate the stability of the closed loop system of the known transfer function in open loop and in particular its polar plot or Nyquist plot. The Nyquist plot of a closed system in feedback is a representation in the Gaussian plane of the value of frequency response function in open loop, $G(j\omega)$, in terms of the real part and imaginary part, with the variation of the pulsation ω.

9.2.9 Digital CMOS Circuits

It is virtually impossible to find electronic devices in our daily lives that do not contain digital circuits, and with most of them having CMOS logic devices at their heart. There are a large number of CMOS devices, divided in a number of families. Using only very few components, it is possible to build fairly elaborate pulse and signal generators.

Depending on the doping material used, there are mainly two types of metal-oxide-semiconductor field-effect transistors (MOSFETs): the *n*-channel, or nMOS, and *p*-channel, or pMOS. In *N*-type metal-oxide-semiconductor (NMOS) logic, *n*-type MOSFETs are used to implement logic gates and other digital circuits. These circuits are mostly used for switching due to their high speed nature, whereas PMOS circuits are slow to transition from high to low state, and their asymmetric input logic levels makes them susceptible to noise. However, metal-oxide-semiconductor (CMOS) technology offers some attractive practical advantages over NMOS technology: high noise immunity and low static-power consumption. CMOS uses a combination of *p*- and *n*-channel MOSFETs as building blocks, but here both low-to-high and high-to-low output transitions are fast since the pull-up transistors have low resistance when switched on, unlike the load resistors in NMOS logic. In addition, the output signal swings the full voltage between the low and high rails. This strong, nearly symmetric response also makes CMOS more resistant to noise [31, 32].

In NMOS circuits the logic functions are realized by arrangements of NMOS transistors, combined with a pull-up device that acts as a resistor. The concept of CMOS circuits is based on replacing the pull-up device with a pull-up network (PUN) that is built using PMOS transistors, such that the functions realized by the PDN and PUN networks are complements of each other. Then a logic circuit, such as a typical logic gate, is implemented as indicated in Fig. 9.24. It comprises a network of NMOS transistors (pull-down network) and a network of PMOS transistors (pull-up network), but each consisting of an equal number of transistors. Each input variable requires an NMOS transistor in the pull-down network and a PMOS transistor in the pull-up network.

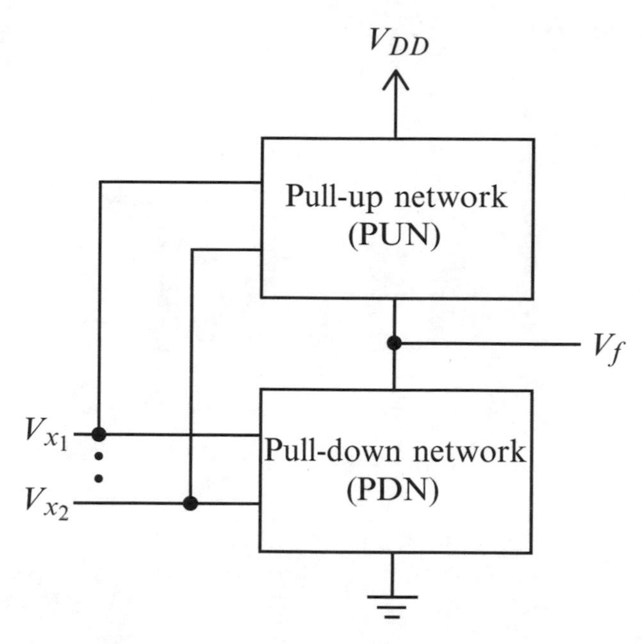

Fig. 9.24 The structure of CMOS circuit

9.2.10 CMOS Inverter

A CMOS inverter (Fig. 9.25) is composed of two MOSFETs, with their gates connected to the inverters input line, and their drains to the output line. The input resistance of the CMOS inverter is extremely high, as the gate of an MOS transistor is virtually a perfect insulator and draws no input direct current. Since the input node of the inverter only connects to the transistor gates, the steady-state input current is nearly zero. A single inverter can theoretically drive an infinite number of gates (or have an infinite fan-out). A CMOS inverter dissipates a negligible amount of power during steady-state operation, which occurs only during switching. This makes CMOS technology useable in low-power and high-density applications.

9.2.11 Current Mirror

Current mirrors replicate the input current of a current sink or current source in an output current, which may be identical or a scaled version. Current mirrors are used to provide bias currents and active loads to circuits. Apart from some special circumstances, a current mirror (Figs. 9.26 and 9.27) is one of the basic building blocks of the operational amplifier; it is a circuit designed to keep the output current

Fig. 9.25 CMOS inverter

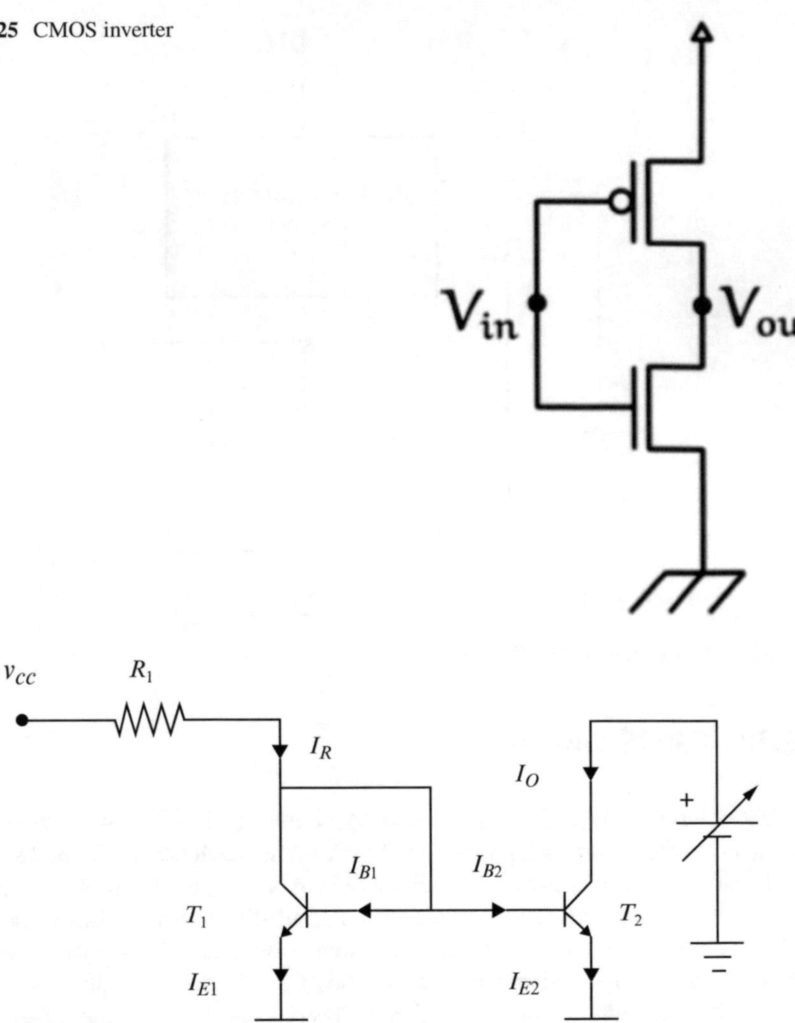

Fig. 9.26 Current mirror with BJT

constant regardless of loading. This type of topology may therefore be used in order to create current generators. The main requirements a current mirror must meet are:

- Output current independence of the output voltage.
- Wide range of output voltages at which the mirror is working properly.
- Low input voltage.

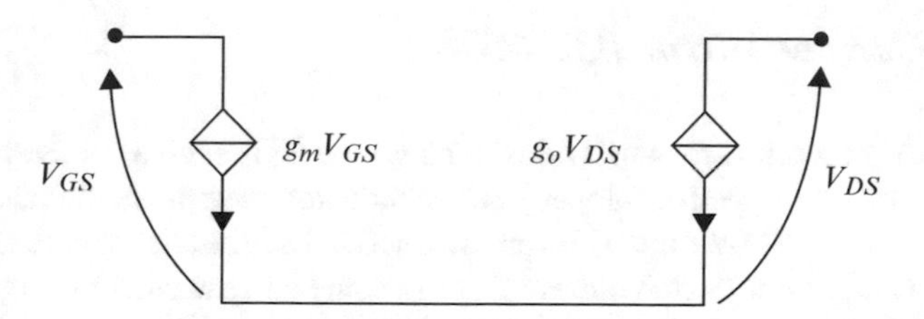

Fig. 9.27 Current mirror with Mosfet (model)

The range of voltages within which the mirror works is called the "compliance range," and the voltage marking the behavior in active/linear region is called the "compliance voltage." There are also a number of secondary performance issues with mirrors, such as temperature stability for example.

9.2.12 Ideal Current Mirror

A current mirror is usually and simply approximated by an ideal current source. However, an ideal current source is unrealistic for several reasons:

- It has an infinite AC impedance, whereas a practical mirror has a finite impedance;
- It provides the same current regardless of voltage, that is, there are no compliance range requirements;
- It has no frequency limitations, whilst a real mirror has limitations due to the parasitic capacitances of the transistors;
- The ideal source has no sensitivity to real-world effects like noise, power supply voltage variations, and component tolerances.

The main part of the current mirror is a bipolar transistors (BJT) or a MOSFET. Transistors in a current mirror circuit must be maintained at the same temperature for precise operation. There are different types of current mirrors:

- Simple current mirror (BJT and MOSFET);
- Base current corrected simple current mirror;
- Widlar current source;
- Wilson current mirror (BJT and MOSFET);
- Cascode current mirror (BJT and MOSFET).

All of the circuits have compliance voltage that is the minimum output voltage required to maintain correct circuit operation: the BJT should be in the active/linear region and the MOSFET should be in the active/saturation region.

9.2.13 Current Mirror BJT/MOS

Current mirror circuits are usually designed with a BJT, such as an NPN transistor, where a positively doped (*P*-doped) semiconductor base is sandwiched between two negatively doped (*N*-doped) layers of silicon. These transistors are specifically designed to amplify or switch current flow. In some current mirror design specifications, the NPN transistor works as an inverting current amplifier, which reverses the current direction, or it can regulate a varying pulse current through amplification to create output mirror properties. One of the reasons that BJTs are used for current mirror design is due to the base-emitter (or PN part) of the transistor functioning reliably like a diode. Diodes regulate both the amount of current that passes and the forward voltage drop for that current. The basic current mirror can also be implemented using MOSFET transistors (Fig. 9.28). In Fig. 9.28, M_1 is operating in the saturation or active mode, and so is M_2. In this setup, the output current I_{OUT} is directly related to I_{REF}. The drain current of a MOSFET I_D is a function of both the gate-source voltage and the drain-to-gate voltage of the MOSFET given by a relationship derived from the functionality of the MOSFET. In the case of transistor M_1 of the mirror, $I_D = I_{REF}$. Reference current (I_{REF}) is a known current and can be provided by a resistor or by a "threshold-referenced" or "self-biased" current source to ensure that it is constant and independent of voltage supply variations.

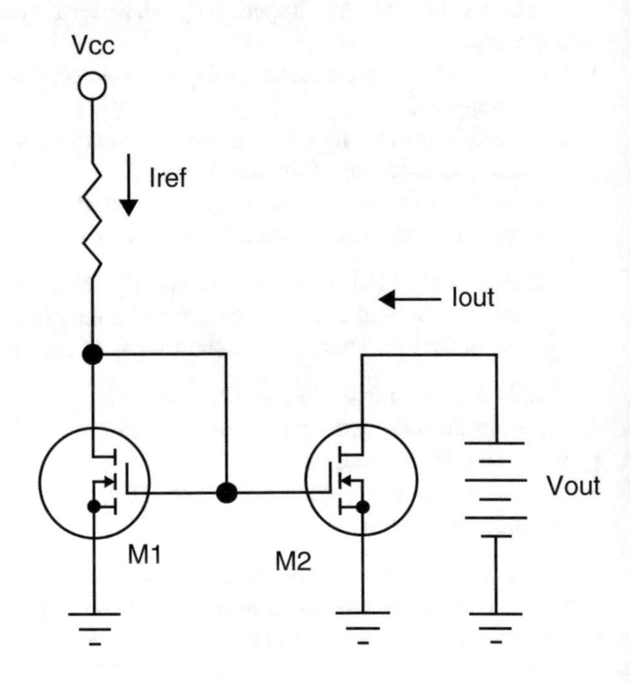

Fig. 9.28 Current mirror with Mosfet

9.3 Low-Power MOSFET

A field effect transistor (FET) works as a conduction channel in the semiconductor with two ohmic contacts, drain and source, where the number of charge carriers in the channel is controlled by a third contact, the gate. The most important FET is the MOSFET. In a MOSFET of silicon, the gate is separated from the channel by an insulating layer of silicon dioxide (SiO_2). The charge carriers of the conductive channel constitute a charge inversion, i.e., electrons in the case of a p-type substrate (n-channel device) or holes in the case of an n-type substrate (p-channel device), that are induced in silicon-insulator interface by the voltage applied to the gate electrode. The MOSFETs are used both as discrete components that active elements in digital and analog monolithic integrated circuits (IC).

9.3.1 General Characteristics of a MOSFET

The structure of a traditional metal-oxide-semiconductor (MOS) is obtained by growing a layer of silicon dioxide (SiO_2) on top of a silicon substrate, depositing in turn a layer of metal or polycrystalline silicon. Since the silicon dioxide is a dielectric material, its structure is equivalent to a planar capacitor, with one of the electrodes replaced by a semiconductor. A transistor of metal-oxide-semiconductor field effect (MOSFET) is based on the modulation of charge concentration by MOS capacitor between an electrode and the gate located over the substrate (body) and isolated from all other regions of the device to a dielectric layer which, in the case of a MOSFET, is an oxide. Compared to the MOS capacitor, the MOSFET includes two additional terminals (source and drain), each is connected to the single highly doped regions and separated from the body region or substrate (substrate). Such regions may be of the type p or n. If the MOSFET is an n-channel or nMOSFET, then the source and drain are "$n+$" and the substrate is a region "p". If the MOSFET is a p-channel or pMOSFET, then the source and drain regions are "$p+$" and the substrate is a region "n". By applying a sufficiently positive voltage between gate and source, the electrons are attracted towards the Si-SiO_2 interface under the gate, forming a conductive channel between source (S) and drain (D). In these conditions, if a voltage is applied $V_{ds} > 0$, a current will flow between drain and source controlled by the voltage V_{gs} between gate and source, that controls the formation of the channel. The electrical characteristics of the MOSFET depend on L (gate length) and W (gate width), as well as the technological parameters such as oxide thickness and doping of the body. Typical values of L and W are: $\mathrm{L} = 0.1 - 2\,\mathrm{pm}$, $\mathrm{W} = 0.5 - 500\,\mu\mathrm{m}$. The range of the gate oxide thickness is 3–50 nm. There are four types of MOS transistors: the N-channel MOSFET (nMOS) is formed on a p-type substrate: nMOS enrichment (enhancement) or normally OFF and nMOS depletion (exhaustion) or normally on. The p-channel MOSFET (pMOS), instead,

Fig. 9.29 Transconductance gm

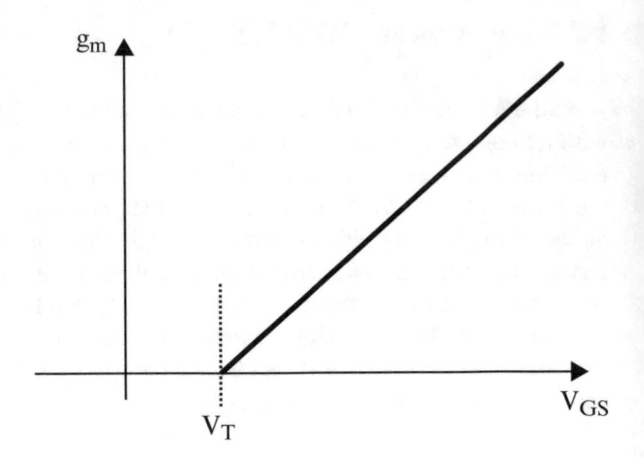

is constructed on an n-type substrate: pMOS enrichment (enhancement) or normally off and pMOS depletion (exhaustion) or normally on.

The operation of the MOSFET Enhancement, or e-MOSFET, can be described using its characteristic curves described above. When the input voltage (V_{in}) (gate) is zero, the MOSFET does not conduct and the output voltage is equal to the supply voltage. In this way, the MOSFET is "fully-OFF" and is in its cutoff region. When input voltage is HIGH or equal to supply voltage (V_{dd}), the operating point of the MOSFET moves to the point A along the load line. The drain current I_D increases to its maximum value due to reduction of the channel resistance, and becomes a constant value independent from V_{dd} and depending only on V_{GS}. Therefore, the device behaves as a closed switch with a minimum channel resistance (R_{ds}). Similarly, when V_{IN} is low or reduced to zero, the operating point of the MOSFET moves from point A to point B along the load line. The channel resistance is very high so that the transistor behaves like an open circuit and no current flows through the channel. If the gate voltage of the MOSFET alternates between two values, high and low, the MOSFET behaves like a sensor "single-throw single-pole" (SPST) in the solid state. Referring to the $I_d - Vd$ characteristic of a MOSFET is also possible to define a set of parameters. While the transconductance gm (Fig. 9.29) gives the variations of the current I_{ds} (output) as a function of the voltage V_{gs}, there is another fundamental parameter that takes the name of output conductance. It expresses the variation of the current I_{ds} as a function of voltage V_{ds}. All two parameters are used to characterize the MOSFET in the amplifier configuration. Other parameters that characterize a MOSFET are the following: R_{ds} (on), the minimum resistance in conduction mode that can vary from a few ohms to a few milliohms; V_{gs} (th) or V_T, the voltage applied to the gate to conduct the Mosfet, is usually greater than 4 V; Q_{GD}, the minimum amount of energy necessary to the gate to switch-on the MOSFET; t_d (on) is the time taken to charge the input capacitance of the device before drain current conduction can start; t_d (off) is the time taken to discharge the capacitance after the after is switched off [33–35].

9.3.2 Mosfet Power Control

By reason of the very high gate resistance (input), its high switching speed and ease of manage, the MOSFET is ideal for layout with operational amplifiers or standard logic gates. In this case, the input voltage of the gate-source should be chosen properly, the device must have a low value of R_{ds} (on) in proportion to the input voltage. The power MOSFETs can be used to control the movement of the DC motors or stepper motors brushless directly from the logic computer or by using the PWM modulation. As a DC motor offers high starting torque proportional to the current, the MOSFET in PWM mode can be used as a speed regulator which provides a smoother and quieter operation of the engine. Since the engine load is inductive, a simple diode is connected across the load to dissipate any electromotive force generated by the motor when the MOSFET is in the "OFF" state. A network locking formed by a zener diode in series with the diode can also be used to permit faster switching and an improved control of the peak inverse voltage and the drop-out time (Fig. 9.30).

9.3.3 Stage of Amplification

The main stages of amplifiers using Mosfet can be described in three configurations: common source, common drain, and common gate. In a common-source amplifier, for example, the input signal is the signal applied to the gate, extracted in output from the drain as current or voltage with respect to ground. The stage

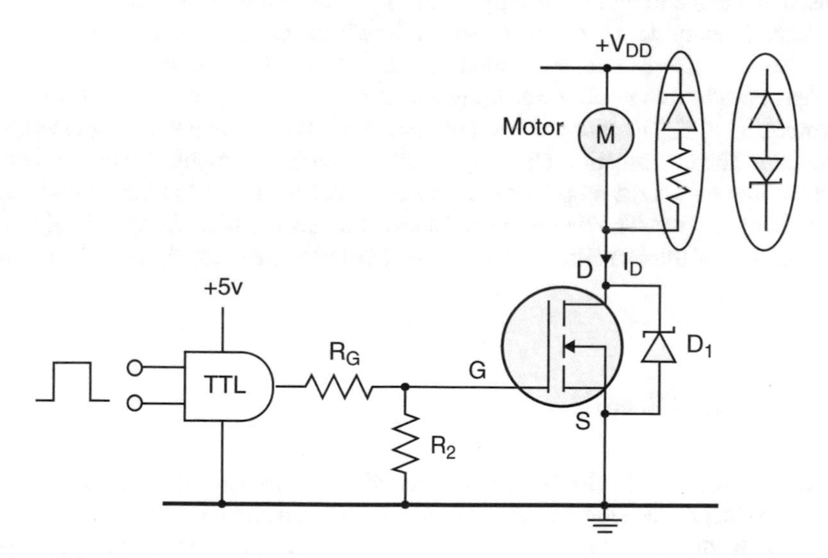

Fig. 9.30 Example of layout for power control

common-source has input resistance very high and aids in the functioning as transconductor and fits perfectly also as a voltage amplifier. This simply means that by reason of high output resistance, the voltage gain (the voltage ratio of the output signal and the input level) depends on the load resistance. This dependence on the load resistance limits its usefulness as a voltage generic amplifier. In a common drain or source amplifier is applied the input signal with respect to ground on the gate terminal of the MOSFET. The main application is related to the buffer, because its input resistance is extremely high, while its output resistance is reasonably low. Unlike a voltage buffer ideal, a source follower MOSFET (common-drain amplifier) provides a gain that is always less than one. Although the common source is capable of substantial power gains of the signal, the voltage gain limits its utility in the application of small signal. The input port of a common gate amplifier has a relatively small input resistance and applies the input signal in current. The output response to the input current applied is significantly extracted as a current signal and the gain is always less than unity. The common gate amplifier can be represented as the dual tracker source or source follower.

9.3.4 Common Source

A common-source amplifier is a topology typically used as a voltage amplifier or transconductance. As transconductance amplifier, the input voltage is seen as a modulation of the current to the load. As a voltage amplifier, instead, the input voltage modulates the amount of current that flows through the MOSFET. However, the output resistance of the device is not high enough for a reasonable transconductance amplifier (ideally infinite), or low enough for a discrete voltage amplifier (ideally zero). Another serious drawback is the limited high frequency response characteristic of the amplifier. Therefore, the output is often routed through both a voltage follower (common-drain stage, C_D), or a current follower (common gate, C_G) to obtain the output characteristics and frequency response more favorable. The combination $C_S - C_G$ is called a cascode amplifier. The bandwidth of the common source amplifier tends to be low, because of the high capacitance due to Miller effect. The limitation on bandwidth in this circuit derives mainly from the coupling of the parasitic capacitance C_{gd} between gate and drain and the series resistance R_A.

9.4 Analog Circuits

An ideal amplifier is an electronic circuit which generates an output signal $y(t)$ given by the product of the signal input $x(t)$ and a constant A. Since both $x(t)$ and $y(t)$ are electrical quantities (voltages or currents), an amplifier can be considered a two-port device: to the input port of a voltage or current signal is applied (possibly without

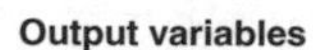

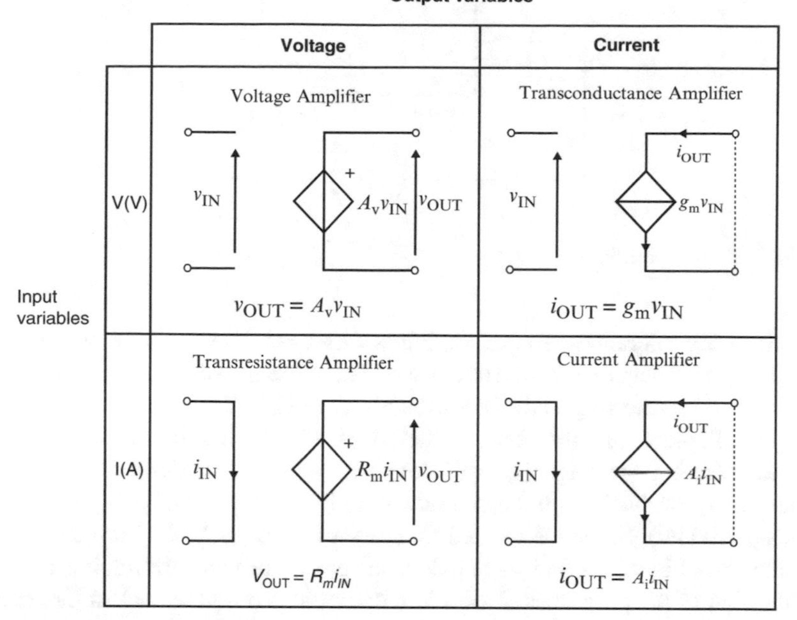

Fig. 9.31 Types of amplifiers

disrupting the circuit that has generated such a signal), while the output port, the amplifier forces a voltage or a current proportional to the input signal by a constant amplification. By depending on the nature (voltage or current), the signals of these amplifiers are classified into four categories as shown in Fig. 9.31.

Contrary to the ideals amplifiers, real circuits have input and finite output resistance. This means that the input port of the actual voltage and transconductance amplifiers is not designed as an open circuit and that the input port of the transresistance amplifiers and current is not a short circuit. The actual amplifiers work correctly only if the generated output signal (voltage or current) takes a value within a specific range known as the amplifier output dynamic. It is also noted that even in the case in which the signal applied to the amplifier input is included inside its input dynamic range, the signal generated at the output port of a real amplifier may not be able to drive properly the connected load by providing to a further limitation of the output dynamics. This limitation does not only depend on the desired output signal, but mainly by the value of the load. In addition to the limitation of bandwidth, the actual amplifiers are affected by a limitation that concerns the maximum slew rate, namely the maximum value of the time derivative of the output voltage. The slew rate limitation is manifested when the output of an amplifier is no longer able to chase the correct output voltage because it changes too quickly. The electronic amplifiers based on semiconductor devices (transistors) are heavily affected by non-idealities and their parameters are poorly controlled due to

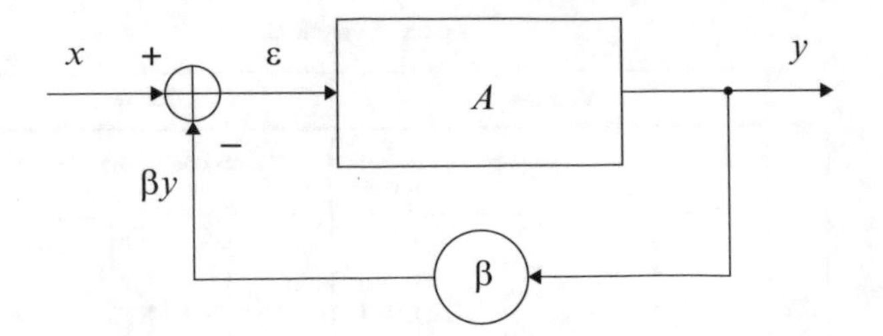

Fig. 9.32 Negative feedback

manufacturing tolerances. To get amplifiers with a behavior comes close to ideality the principle of negative feedback is used. The principle of negative feedback can be described by referring to the block diagram of Fig 9.32.

In this diagram, an amplifier (characterized by a high gain) is used to amplify the difference between signals, that is, the error between the desired signal and the output signal, set back from the entrance to block β (a passive block, for example, a voltage divider). Since we wanted to a negative feedback, the correction should vary the output in such a way as to reduce the error. For this purpose, it is important that the sign of the proportional term in output that is reported back at the entrance is negative. By referring to the block diagram of Fig. 9.32, the output signal y can be expressed as follows:

$$y = A(x - \beta y) \tag{9.19}$$

$$y = \frac{A}{1 + A\beta} x = \frac{1}{\beta} \frac{A\beta}{1 + A\beta} x \tag{9.20}$$

It may also observe that the error signal ϵ, at the entrance of the block A, is given by the following equation:

$$\epsilon = x - \beta y = \frac{1}{1 + A\beta} x \tag{9.21}$$

9.5 Operational Amplifier

The operating principle of negative feedback can be used advantageously to design highly accurate amplifiers. To be able to reconstruct the block diagram of Fig. 9.32, first of all, is needed a differential amplifier, i.e., a block that amplifies the difference between two input quantities (typically voltage) so as to realize the different block. In addition it requires an amplifier with a high (it is not necessary, however, to be accurate) differential amplification (A), so that the gain of $A\beta$ is much greater than 1. Differential amplifiers are one of the most important analog circuits known as operational amplifier or, more simply, as the op-amp (Fig. 9.33).

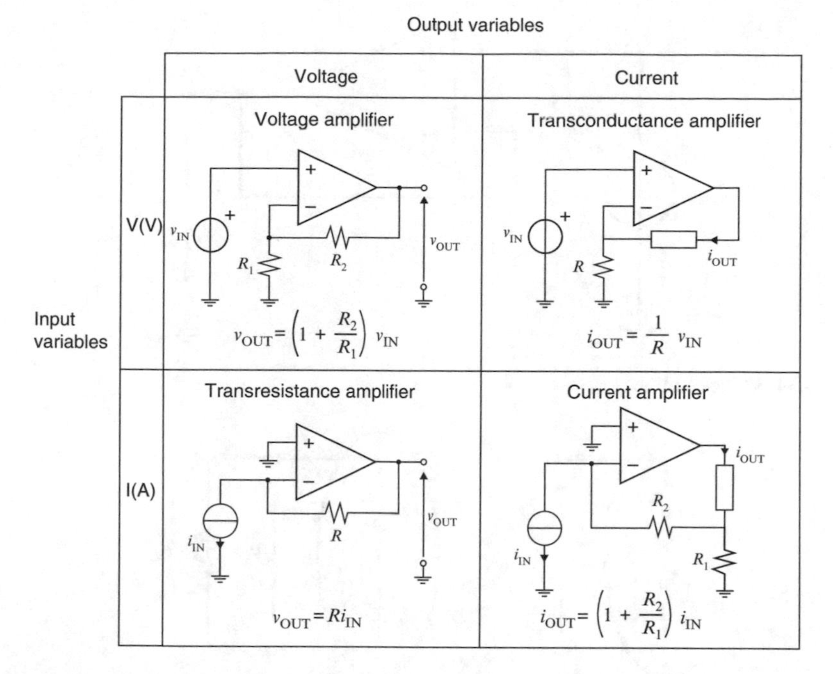

Fig. 9.33 Circuit diagrams with operational amplifiers

A simple and important circuit based on the use of op-amps is shown in Fig 9.34. Such a circuit is known as a voltage follower, voltage buffer, or voltage follower: that is, an ideal voltage amplifier with amplification equal to one. The voltage follower importance is linked to the concept of load. If you want to generate a specific voltage across a load connected to a port of a circuit, then to connect the load as shown at the top of Fig. 9.35 is typically not possible, because it is the load introduces a disturb effect on the voltage. However, by connecting a voltage follower with the operational amplifier between the source and the load as shown in the lower part of Fig. 9.35, the voltage source is not disrupted because the absorbed current from the non-inverting op-amp is nothing with output voltage $V_{out} = V_{IN}$.

9.6 Power Supply and Rejection

In the last paragraphs we have described the operational amplifiers like any other electronic circuit, it requires a supply voltage continuously in order to work. Although it is permissible to assume that these voltages are constant, often they are actually disturbed with residual oscillations which should not influence the output voltage of an op-amp, however, the output voltage of an actual operational amplifier in practice shows an error proportional to the fluctuations of the supply voltage.

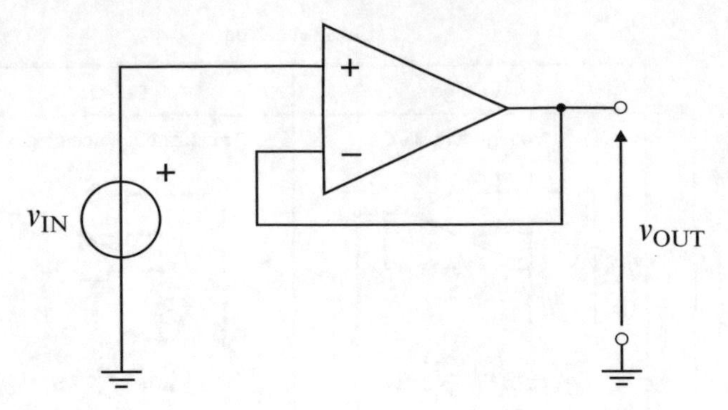

Fig. 9.34 Voltage follower

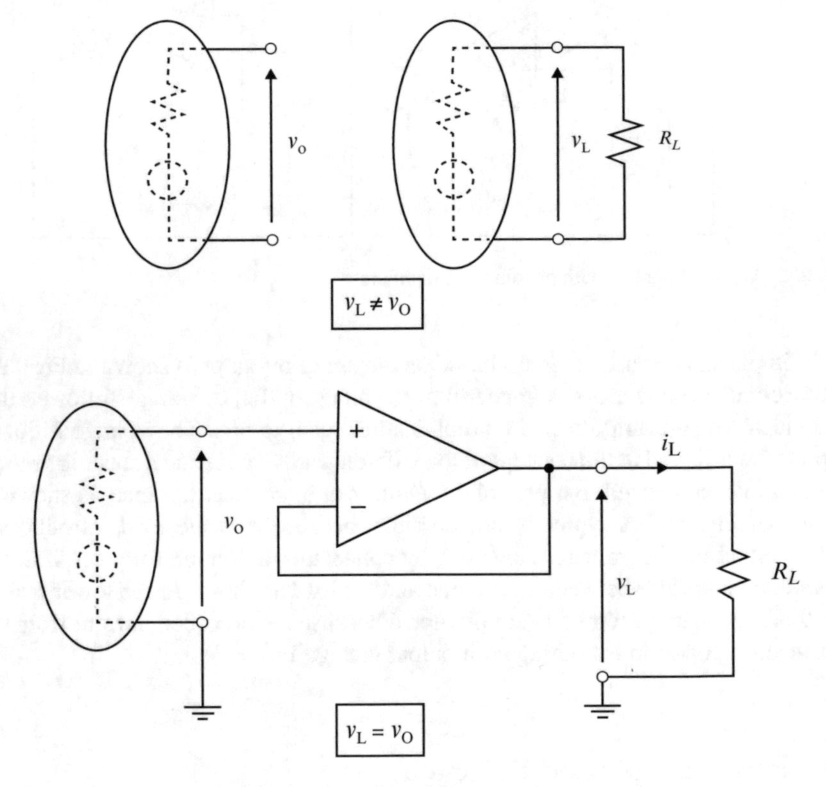

Fig. 9.35 Impedance mismatch with the voltage follower

The performance of an operational amplifier is often expressed in terms of rejection of the supply voltage ratio (power supply rejection ratio), PSRR, defined as:

$$\mathrm{PSRR} = \frac{A}{A_{\mathrm{ps}}} \tag{9.22}$$

$$A = \frac{V_{\text{out}}}{V_d} \tag{9.23}$$

$$A_{\text{ps}} = \frac{V_{\text{out}}}{V_{\text{ps}}} \tag{9.24}$$

Since both A and A_{ps} depend on the frequency, the PSRR of an operational amplifier depends on the frequency. At low frequencies, PSRR can have values greater than 120 dB, whereas at higher frequencies (order of 100 kHz) the PSRR could be about 30–40 dB.

9.7 Low Noise Pre-amplifiers

Low noise amplifier (LNA) is an electronic amplifier used to amplify possibly very weak signals. A pre-amplifier is an electronic device which amplifies an analog signal. Generally is the stage which anticipates the high power amplifier.

In this example we design a simple pre-amplifier with background noise of the order of 0.8 nV/Hz @ 1 kHz. The circuit diagram is shown in Fig. 9.36. The main components used are LT1128 dell Linear Tech and the JFET IF3602 of InterfetCorporation. LT1128 (Figs. 9.37 and 9.38) is an operational amplifier ultra noise at high speed. Main characteristics are the following:

- Noise voltage: 0.85 nV/Hz @ 1 kHz
- Bandwidth: 13 MHz
- Slew rate: 5 V/uS
- Offset voltage: 40 uV.

The IF3602, instead, is a Dual-N JFET used as stage for input of the operational amplifier.

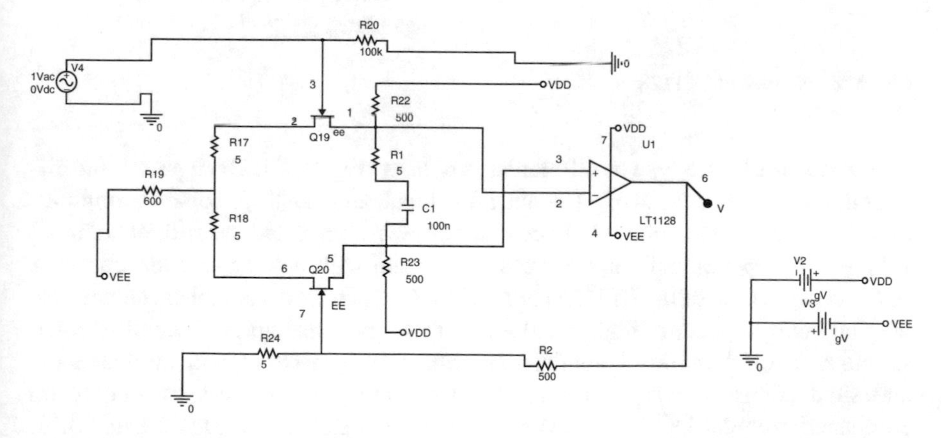

Fig. 9.36 Low noise pre-amplifier

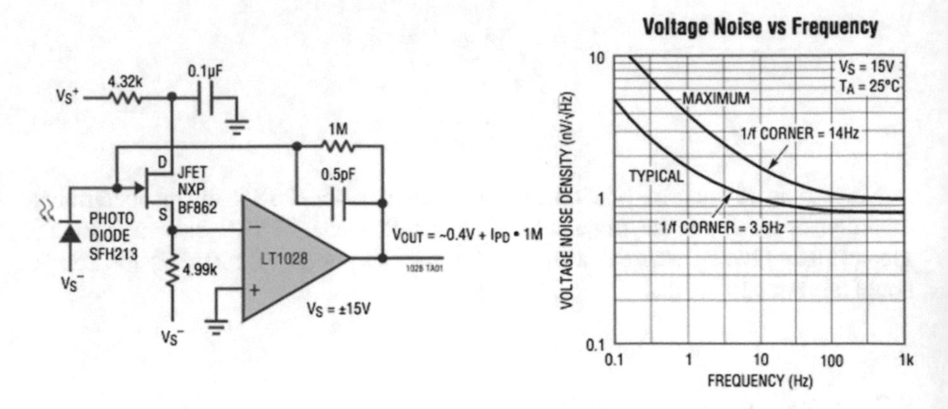

Fig. 9.37 Voltage noise LT1128

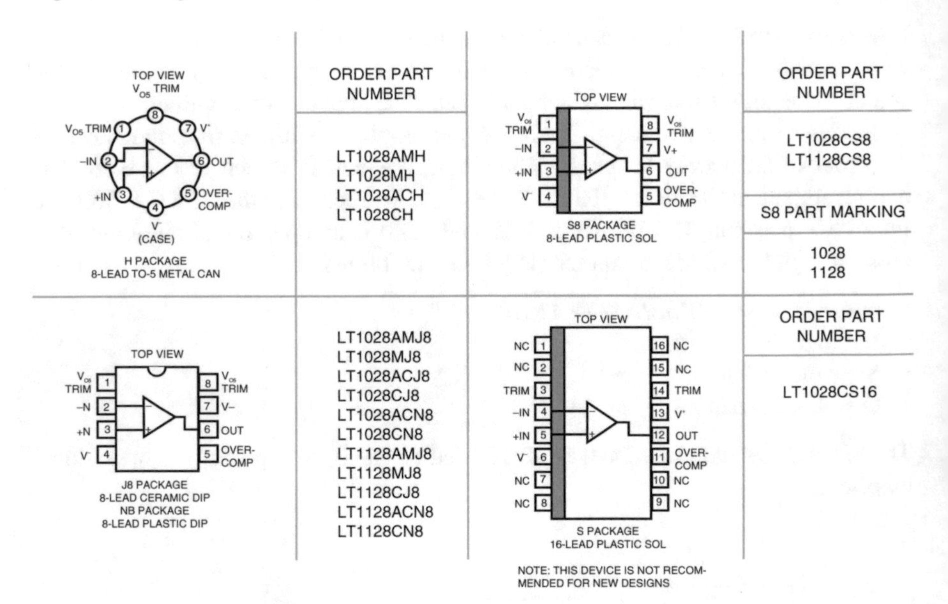

Fig. 9.38 Pin-outs of LT1128

The op-amp is one type of differential amplifier (Fig. 9.39). The inputs of the differential amplifiers consist of a $V+$ and a $V-$ input, and ideally the op-amp amplifies only the difference in voltage between the two, which is called the differential input voltage. The operational amplifier can be realized with bipolar junction transistor (BJT, as in the case of the LT1128) or MOSFET, which works at higher frequencies, with an input impedance higher and a lower energy consumption. The differential structure is used in those applications where it is necessary to eliminate the undesired common components to the two inputs. In this way, in output are eliminated eventual DC components on the input signal such as, the thermal drift.

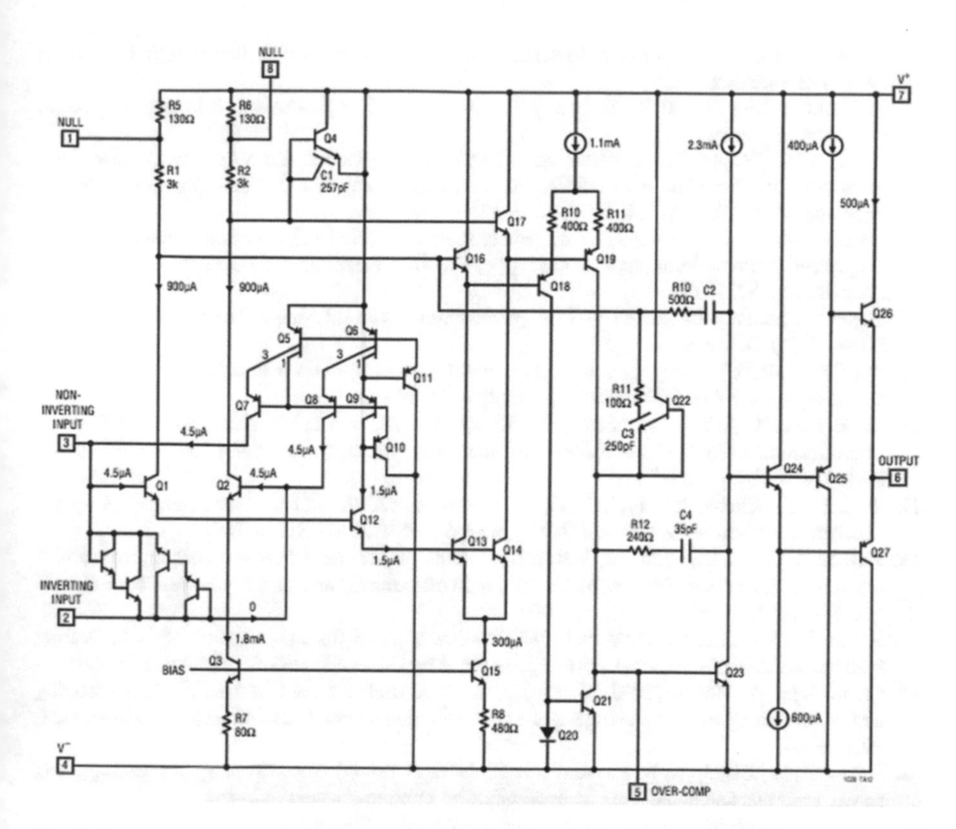

Fig. 9.39 The internal structure of the LT1128

The noises of internal current and voltage of the amplifiers depend on the intrinsic physical phenomena and are by their nature random, aperiodic, and uncorrelated. Typically have a distribution of amplitude of Gaussian. The relationship between the peak to peak value and the effective value of these components is statistical. A possible qualitative rule is that the RMS value multiplied by 6 does not exceed the peak to peak in the 99.73 % of cases.

References

1. Park, J., & Mackqy, S. (2003). *Practical data acquisition for instrumentation and system control*. Oxford: Elsevier.
2. Lacanette, K. (2003). *National temperature sensors handbook*. Handbook Ann. Mat. National semiconcductor.
3. National Instruments (1996). Data acquisition fundamentals, Application Note 007.
4. National Instruments (1996). *Signal conditioning fundamentals for PC-based data acquisition systems*, Handbook National Instruments.

5. Taylor, J. (1986). *Computer-based data acquisition system*. Research Triangle Park: Instrument Society of America.
6. Di Paolo Emilio, M. (2013). *Data acquisition system, from fundamentals to applied design*. New York: Springer.
7. Roundy, S., Wright, P., & Pister, K. (2002). Micro-electrostatic vibration-to- electricity converters. In *Proceedings of ASME International Mechanical Engineering Congress and Exposition IMECE2002* (Vol. 220, pp. 17–22).
8. Stordeur, M., & Stark, I. (1997). Low power thermoelectric generator: Self- sufficient energy supply for micro systems. In *Proceedings of the 16th International Conference on Thermo-Electrics* (pp. 575–577).
9. Shenck, N., & Paradiso, J. (2001). Energy scavenging with shoe-mounted piezoelectrics. *IEEE Micro, 21*(3), 30–42.
10. Roundy, S. (2003). *Energy scavenging for wireless sensor nodes with a focus on vibration to electricity conversion*. PhD thesis, University of California.
11. Tsutsumino, T., Suzuki, Y., Kasagi, N., Kashiwagi, K., & Morizawa, Y. (2006, November). Micro seismic electret generator for energy harvesting. In *Technical Digest PowerMEMS 2006*, Berkeley, USA (pp. 133–136).
12. Sterken, T., Altena, G., Fiorini, P., & Puers, R. (2007, April). *Characterisation of an electrostatic vibration harvester, DTIP of MEMS and MOEMS*, Stresa, Italy.
13. Sterken, T., Baert, K., Puers, R., & Borghs, S. (2002, November). Power extraction from ambient vibration. In *Proc. SeSens. Workshop on semiconductor sensors*, Veldhoven, Netherlands (pp. 680–683).
14. Szarka, G., Stark, B., & Burrow, S. (2012). Review of power management for energy harvesting systems. *IEEE Transactions on Power Systems, 27*(2), 803–815. ISSN: 0885-8993.
15. Cammarano, A., Burrow, S. G., Barton, D. A. W., Carrella, A., & Clare, L. R. (2010). Tuning a resonant energy harvester using a generalized electrical load. *Smart Materials and Structure, 19*, 055003.
16. Guyomar, D., Badel, A., Lefeuvre, E., & Richard, C. (2005). Toward energy harvesting using active materials and conversion improvement by nonlinear processing. *IEEE Transactions on Ultrasonics, Ferroelectrics, and Frequency Control, 52*, 584–595.
17. Mitcheson, P. D., Stoianov, I., & Yeatman, E. M. (2012). Power-extraction circuits for piezoelectric energy harvesters in miniature and low-power applications. *IEEE Transactions on Power Electronics, 27*, 4514–4529.
18. Szarka, G. D., Burrow, S. G., & Stark, B. H. (2012). Ultra-low power, fully-autonomous boost rectifier for electro-magnetic energy harvesters. *IEEE Transactions on Power Electronics, 28*(7), 3353–3362. doi:10.1109/TPEL.2012.2219594.
19. Maurath, D., Becker, P. F., Spreeman, D., Manoli, Y. (2012). Efficient energy harvesting with electromagnetic energy transducers using active low-voltage. *IEEE Journal of Solid-State Circuits, 47*(6), 1369–1380
20. Beeby, S. P., Tudor, M. J., & White, N. M. (2006). Energy harvesting vibration sources for microsystems applications. *Measurement Science and Technology, 17*, R175–R195.
21. Khaligh, A., Zeng, P., & Zheng, C. (2010). Kinetic energy harvesting using piezoelectric and electromagnetic technologies–state of the art. *IEEE Transactions on Industrial Electronics, 57*(3), 850–860.
22. Paulo, J., & Gaspar, P. D. (2010). Review and future trend of energy harvesting methods for portable medical devices. In *Proceedings of the World Congress on Engineering* (Vol. 2).
23. Zhu, D., Tudor, M. J., & Beeby, S. P. (2010). Strategies for increasing the operating frequency range of vibration energy harvesters: A review. *Measurement Science and Technology, 21*, 022001-1–022001-29.
24. Cepnik, C., Lausecker, R., & Wallrabe, U. (2013). Review on electrodynamic energy harvesters – a classification approach. *Micromachines, 4*(2), 168–196. http://www.mdpi.com/2072-666X/4/2/168. Accessed 20 Jan 2015.
25. Ulaby, F. T., Michielssen, E., & Ravaioli, U. (2010). *Fundamentals of applied electromagnetics* (6th ed.). Upper Saddle River: Prentice Hall.

26. Roundy, S., Wright, P. K., & Rabaey, J. M. (2003). A study of low level vibrations as a power source for wireless sensor nodes. *Computer Communications, 26*(11), 1131–1144.
27. Sazonov, E., Li, H., Curry, D., & Pillay, P. (2009). Self-powered sensors for monitoring of highway bridges. *IEEE Sensors Journal, 9*, 1422–1429.
28. Toh, T. T., Mitcheson, P. D., Holmes, A. S., & Yeatman, E. M. (2008). A continuously rotating energy harvester with maximum power point tracking. *Journal of Micromechanics and Microengineering, 18*, 104008-1-7.
29. Howey, D. A., Bansal, A., & Holmes, A. S. (2011). Design and performance of a centimetre-scale shrouded wind turbine for energy harvesting. *Smart Materials and Structures, 20*, 085021.
30. Razavi, B. (2002). *Design of analog CMOS integrated circuits*. New York: McGraw-Hill.
31. Razavi, B. (2008). *Fundamentals of microelectronics*. Chichester: Wiley.
32. Sedra, A. S., & Smith, K. C. (2013). *Microelectronic circuits*. Oxford: Oxford University Press.
33. Razavi, B. (2002). *Design of integrated circuits for optical communications*. New York: McGraw-Hill.
34. Di Paolo Emilio, M. (2015). *Microelectronics from fundamentals to applied design*. Cham: Springer.
35. Di Paolo Emilio, M. (2014). *Embedded system design for high speed data acquisition and control system*. Cham: Springer.

Chapter 10
Low-Power Solutions for Biomedical/Mobile Devices

10.1 Introduction

Sensor networks have internal wireless modules solutions for the transmission of the information to a remote computer and then a smartphone. The sensors with a wireless connection can be supplied autonomously with energy harvesting techniques, by using further battery to extend the life of the system. The worst case scenario in wearable devices is the inability to take advantage of the mechanical energy in patients with immobility in bed for various health reasons, where wearable devices must rely on battery-powered solutions. There is no mechanical energy, the intensity of light is too low to power a low-power device. Therefore only a part of the head and sometimes the area of the wrists represent the only areas relatively suitable for the harvesting of energy: thermal and light. The available low power is connected to the level of internal illumination and the heat transfer from the person that is considerably low, the latter determined by the natural convection around the head. On one side there is provided a device that generates electricity from the internal temperature difference of the human body, create the thermoelectric micro-generators (TEG). They are able to generate a few microwatts of electrical power. A BiCMOS device is made using standard materials to maintain the reduced cost. The low thermal conductivity of these materials appears to be the main factor to increase the output power. The materials used are the poly-Si and poly-SiGe. On the other side, there is a piezoelectric device that converts the kinetic energy of a mass free to move inside of a rigid frame into electrical energy due to the impact of this mass with the two ends of the frame. The real-time monitoring of the patient's vital signs requires the use of wearable sensors and mobile devices [1–15].

M. Di Paolo Emilio, *Microelectronic Circuit Design for Energy Harvesting Systems*, DOI 10.1007/978-3-319-47587-5_10

10.2 Design of Wearable Devices

The goal is to obtain an efficient energy harvesting by reducing the size of the battery. The classical example is typical of pacemakers used in humans. If the harvesting of vibrational energy in this case is realized, it is possible to reduce the size of the pacemakers and to extend the life of the battery. They use internal sensors to monitor the status of a patient, or are implanted to save life. Currently, the size of a typical pacemaker is about 42 mm–51 mm–6 mm. Typically, the battery requires about 60 % of the size of the pacemaker. The goal is to reduce the area and then set the maximum size of the energy harvesting techniques such as 25 mm–25 mm–5 mm with power consumption about 10 uW. Fortunately, the packaging of traditional pacemaker eliminates the need to use biocompatible materials. The batteries and the pacemaker circuits are encapsulated in a titanium case: a biocompatible material with a sealed casing which ensures that there is no contact between the interior of the body and the pacemaker batteries or circuits.

The humans constantly produce heat as a side effect of the metabolism. However, only a part of this heat is dissipated in the environment as a flow of heat and infrared radiation, the remainder is rejected in the form of water vapor. Furthermore, only a small fraction of the heat flow can be used in collection of energy. The heat flux can be converted into electrical energy using a thermoelectric generator (TEG). The body has high thermal resistance; therefore, the heat flow is rather limited (Figs. 10.1 and 10.2). The main problem of these devices is the duration of the battery, even though they have much longevity, the battery must be replaced: the patient must go to hospital, he undergoes a surgical intervening, with possible complications, with a few days of recovery, etc. Here comes the energy harvesting to create a device that is fueled by the energy produced by the human body. In practice this would eliminate the need an intervention for replacement of the battery. The objective can be to develop a micro-generator TEG that is cheap, small, and low-power consumption,

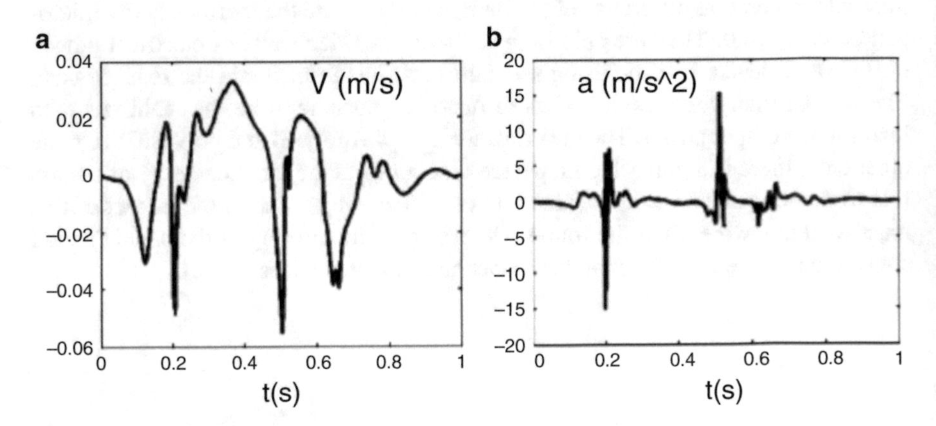

Fig. 10.1 Acceleration profile and heart rate speed [response]

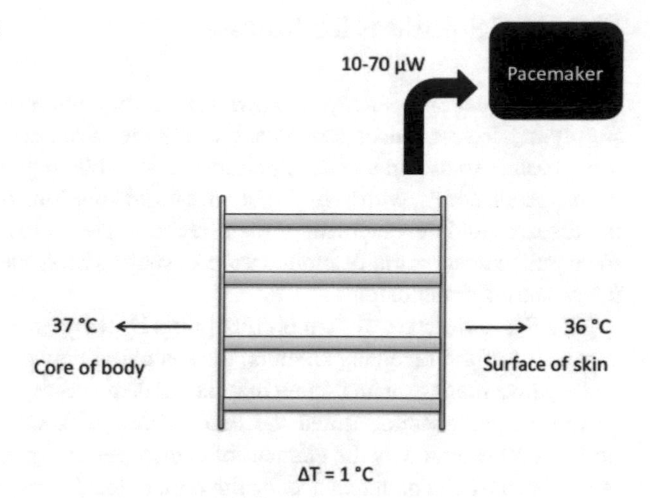

Fig. 10.2 Energy harvesting for a pacemaker

as, for example, materials such as poly-Si and poly-SiGe, in BiCMOS configuration (Bipolar Complementary Metal Oxide Semiconductor), which is a technology that integrates CMOS and BJT on the same semiconductor chip. The advantage of this procedure can be described in the following points:

- Two different technologies.
- Overall capacity of a BiCMOS gate is almost equal to that of the only BJT.
- The possibility to combine in the same integrated analog and digital electronics.

A TEG generator consists of n-type and p-type thermoelectric material. These are electrically connected in series by means of metal bridges and are arranged to use the best available area. The generator layer is positioned between the silicon substrate and the heat stabilizer. A thermal gradient between the lower end of the device and the environment at the upper end is creating a vertical heat flow, which can be partially converted into electrical energy by the effects of thermoelectricity [16–20]. Research into Body Area Network (BAN) and their applications are experiencing growing interest, especially in monitoring the status of the human body health, monitor their movements and activities. Wearable wireless sensors make this possible, by collecting information on the individual or single community, to anticipate their requests before they are expressed. Another possible application is under Ambient-Assisted Living (AAL): by collecting physiological information in real time to prevent hazardous events such as heart attack.

10.3 RF Solutions for Mobile

The radio frequency energy is nowadays a very interesting source for the power supply in wireless sensor networks; through the wireless power distribution systems it is possible to obtain an infrastructure ad hoc able to power networks of hundreds or thousands nodes, with a single source of transmission. Today, the real challenge in the research and development of these technologies is to recover power transmitted from public telecommunications services, such as broadcast TV and radio or mobile telephone communications.

The EH radio wave system is constituted by a system commonly called rectenna composed of the receiving antenna, the matching network, and the rectifier circuit, and a power management system that includes the dc-dc converter, the accumulation system (supercapacitors), and the load (consumer). One drawback to the sensor nodes is represented by the element of energy reserve, generally constituted by batteries, because the maintenance for the replacement of batteries with a large number of nodes is very expensive and also to aggravate the maintenance costs. Therefore, there is need to make autonomous from an energetic point of view, the many sensor nodes located in the various environments. The solution proposed to lengthen the lifetime of the batteries or completely prevent its use is based precisely on the recovery of energy from environmental sources, particularly from the airwaves. The benefits to these technologies can indirectly arise from recent developments in micro- and nano-technologies, which have led to the creation of ever-smaller devices and with energy consumption which are reduced accordingly. The energy harvesting system compared to traditional power systems is the great saving in implementation and operating costs The rectenna constitutes the transducer of an EH radio frequency system, and is therefore capable of converting the incident RF signal on the antenna into a DC signal to be amplified and accumulated in the power management circuit. The antenna is the transmitter of the radio frequency in the EH system. When an electric field impacts with a certain intensity on it, an induced voltage that is generated must be rectified through a rectifier circuit [21–30].

10.3.1 Ferrite Rod Antenna

As shown in Fig. 10.3, it is a ferrite core with a number of turns wound around it. The purpose is to obtain a sort of small loop antenna (magnetic dipole) around the ferrite core. This antenna, together with electronics that there is downstream, can directly replace the batteries in specific applications, for example, for WSNs. It constitutes a good solution to the low frequencies, in particular, for medium-wave transmissions.

Figure 10.4 shows the antenna equivalent circuit also described in resonance condition; R_R is the radiation resistance of the loop, that is:

Fig. 10.3 Ferrite rod antenna

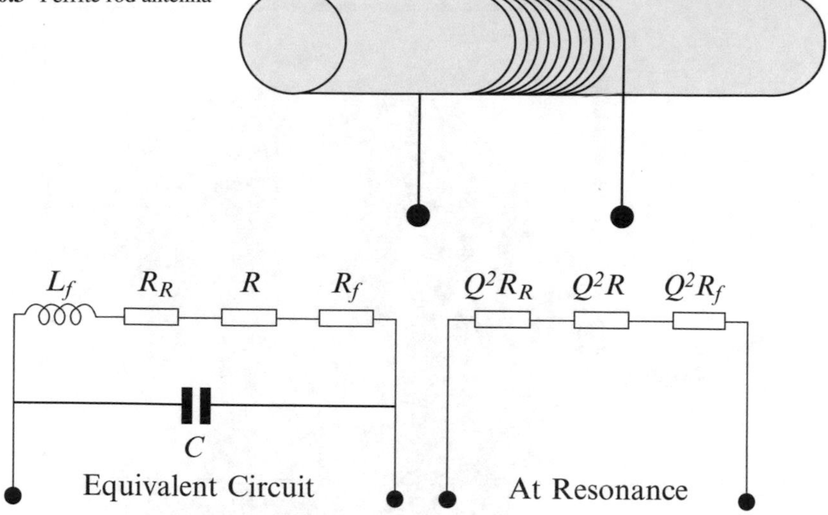

Fig. 10.4 Equivalent circuit of the ferrite rod antenna

$$R_R = 31,200 * \left(\frac{\mu_e nA}{\lambda^2}\right)^2 \tag{10.1}$$

where n is the number of turns, A is the loop section, μ_e is the relative magnetic permeability of the ferrite. R_f is the loss resistance of the rod, that is:

$$R_f = 2\pi f \mu_e \frac{\mu''}{\mu'} \mu_0 n^2 \frac{A_f}{l_f} \tag{10.2}$$

where f is the operating frequency, μ' and μ'' are the real and imaginary parts of the magnetic permeability of the ferrite, respectively, A_f is the cross sectional area of the rod, and l_f the length of the rod. The inductance of this antenna is instead given by:

$$L_f = \mu_e \mu_0 n^2 \frac{A_f}{l_f} \tag{10.3}$$

By using the capacitor C in parallel is possible to convert the antenna impedance, at the resonance frequency f_0, in a pure resistance whose value is greater of the current loop resistance and Q is the quality factor defined as follows:

$$Q = \frac{f_0}{\Delta f} \tag{10.4}$$

where Δf is the bandwidth at -3 dB.

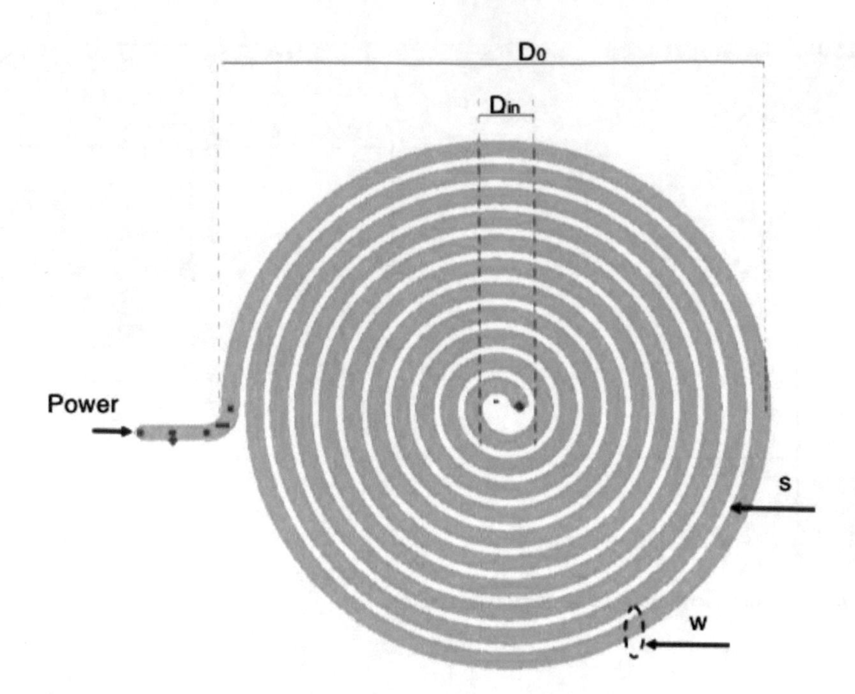

Fig. 10.5 Circular spiral inductor antenna

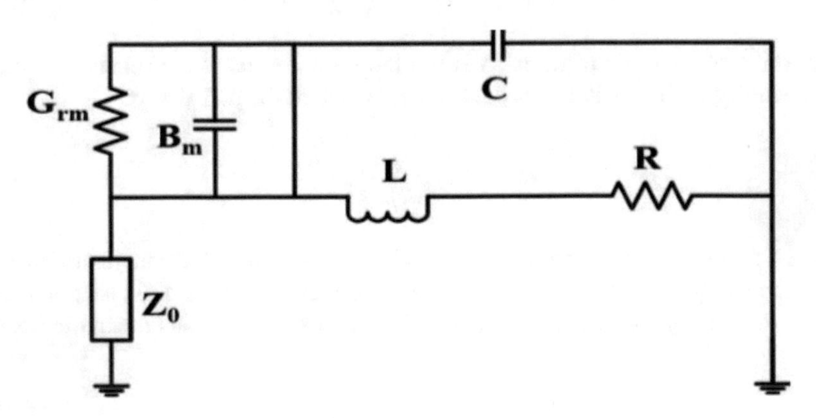

Fig. 10.6 Equivalent circuit of the circular spiral inductor antenna

10.3.2 Circular Spiral Inductor Antenna

The circular spiral antenna (Fig. 10.5) ensures a high power density and this is one of the most used for the recovery of energy from radio waves. This antenna may be supplied by a microstrip line and in this case it has the equivalent circuit of Fig. 10.6.

The total magnetic energy stored in the circular spiral is expressed through an inductance defined as:

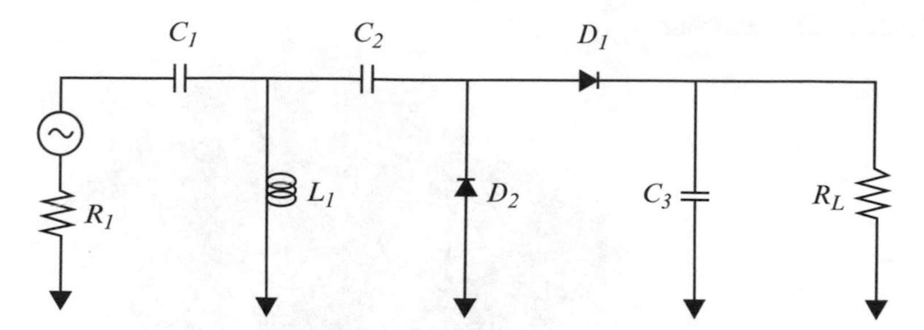

Fig. 10.7 Equivalent circuit of the folded dipole

$$L = \frac{\mu N^2 D_m}{2}\left(\ln\left(\frac{2,46}{\rho}\right) + 0,2(\rho^2)\right) \tag{10.5}$$

where μ is the magnetic permeability of the material, ρ is the fill-ratio, N the number of windings, and $D_m$the mean diameter of the inductor.

10.3.3 Folded Dipole

The folded dipole, with respect to a common dipole $\lambda/2$, allows to have a radiation resistance (approximately 300 Ω for a folded dipole two-wire), a voltage V_{out}, and a higher operating band. It is usually coupled with a voltage multiplier rectifier to further increase the value of the dc voltage available.

The power density is the most important parameter for the RF energy harvesting; the expected level of the received power depends on level of transmitted power and the distance from the transmitter, while it appears to be independent of frequency (Fig. 10.7).

10.3.4 Microstrip Antenna

The microstrip antenna (patch antenna) is the common antenna used for modern cell (Figs. 10.8 and 10.9); it works well at high frequencies but has narrow band, so that even small frequency deviations may cause large losses of the recovered power. As the rectifier circuit can be used a single Schottky diode which allows to have a higher conversion efficiency [31].

There are various configurations of microstrip antennas (L-shape, E-shape, and U-shape), all of narrow-band type (narrow band). In Fig. 10.9 is shown the equivalent circuit of a microstrip antenna composed of two inductors and a shunt capacitance.

Fig. 10.8 Microstrip antenna

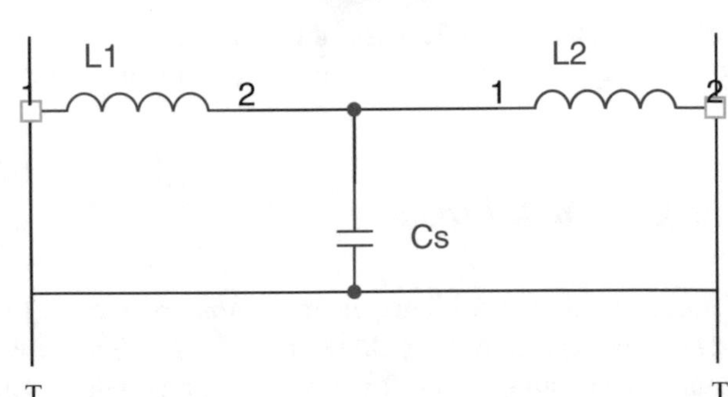

Fig. 10.9 Equivalent circuit of microstrip antenna

10.4 Power Management

The sector of the energy harvesting recently had a growing interest, and therefore many studies have been directed towards the design of integrated circuits for converters. In particular, the company of "Linear Technology" has recently produced a wide range of ultra-low power integrated circuits for applications of EH. These circuits may primarily to convert vibrational energy (piezo), solar and thermal (thermopile, thermocouple) and have a high conversion efficiency. Some of these may also be exploited for the radio wave energy conversion. These ICs are particularly relevant for applications in industrial automation, transportation, automotive, and wireless sensors. The LTC3108 of Linear Technology is a boost converter with high efficiency which can operate with very low input voltages, ranging from 20 mV to 0.5 V, and therefore can be used for energy harvesting system in radio frequency.

The integrated LTC3108 can also work as AC Energy Harvester due to internal rectifier circuit; in such case, it is possible to connect the source directly to the circuit. The expected pattern is the following (Fig. 10.10), in which the input is

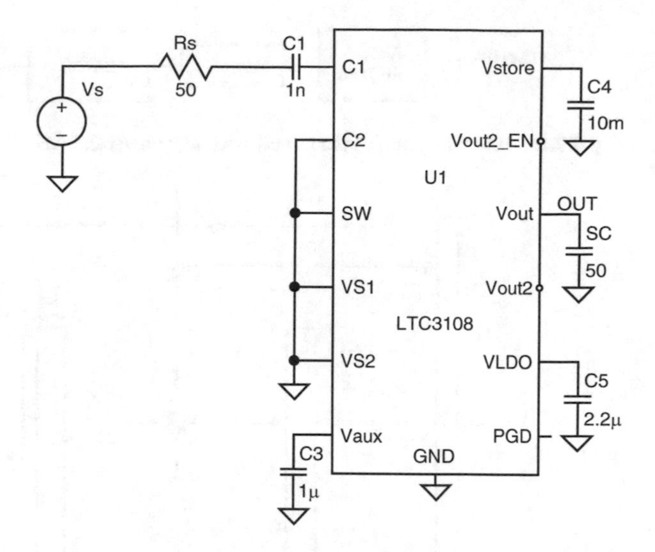

Fig. 10.10 AC energy harvester

an antenna designed with a sinusoidal generator $V_s = 225\,\text{mV}$ at 102 MHz and a resistance R_s in series from 50 Ω; C_2 and SW pins are shorted to ground because it is not expected the presence of the transformer.

The LTC3105 of Linear Technology is a high efficiency boost converter that can operate with input voltages ranging from 225 mV to 5 V. It has integrated a maximum power point controller (MPPC), which maximizes the energy that can be extracted from any source, and a voltage regulator dissipative Low Drop Out (LDO).

10.5 Ultra-Low Power 2.4 GHz RF Energy Harvesting and Storage System

An RF energy harvesting system optimizes energy management for many devices. The circuit described in this paragraph is optimized for the power range typically observed in the ISM 2.4 GHz band in order to collect the energy supplied by the nearby Wi-Fi, Bluetooth, and other devices. In this layout, the rectified voltages are low, then the operation of the power management circuit in 100 mV scheme is critical. The RF energy harvesting system proposed can be thought in terms of four main blocks, as shown in Fig. 10.11. A key feature is the boost converter that represents the start to ultra-low input voltages, while maintaining high efficiency to useful output voltages. Finally, we need a reserve of energy like a battery, large capacitor, or supercapacitor to store energy from the RF input sources until sufficient energy is available for a sensing operation [32–35].

The antenna is one of the most important components of an RF energy harvesting system. It collects ambient RF power and transmits it to the rectifier; so it must be designed to capture more energy. In Fig. 10.12 an example of a patch antenna

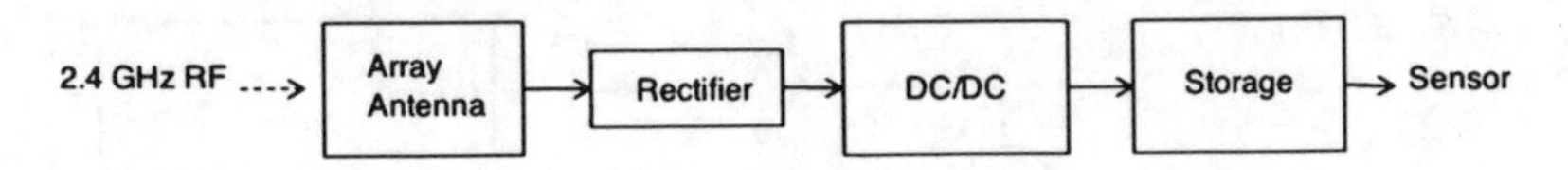

Fig. 10.11 Block diagram of the circuit ultra-low power 2.4 GHz

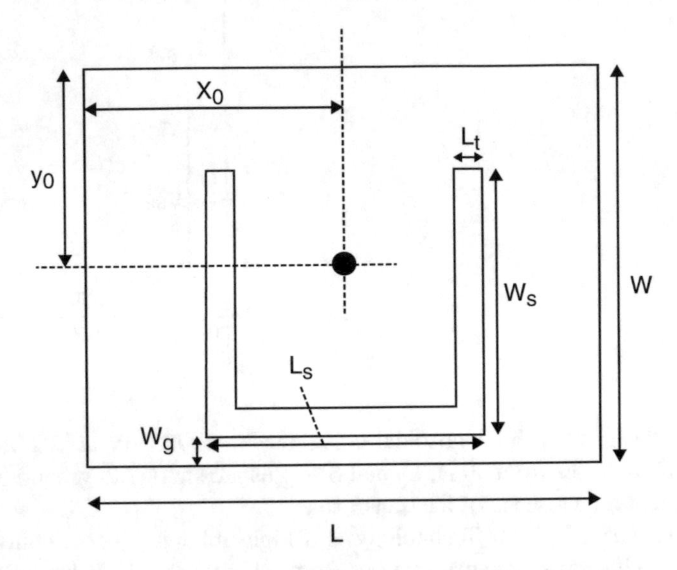

Fig. 10.12 Antenna

consists of a U-groove structure. The antenna is printed on a substrate with a thickness of 6 mm and a dielectric constant of 10.2. The radiating element has dimensions of $L = 0.155\lambda = 19\,\text{mm} \times W = 0.118\lambda = 14.5\,\text{mm}$. $L_s = 10.15\,\text{mm}$, $L_t = 1\,\text{mm}$, $W_g = 1\,\text{mm}$, $x_0 = L/2$, $Y_0 = W/2$. The dimensions of the ground plane are $0.326\lambda \times 0.326\lambda$ (40 mm × 40 mm).

The rectifier is a single phase voltage doubler that uses the detector diodes Avago HSMS-286C with the microstrip RF matching networks on the substrate. A typical DC-DC boost converter connected an input capacitor through an inductor in an output capacitor under the control of a transistor. High efficiency is possible with a low-loss transistor with the inductor size and switching frequency of the transistor carefully selected on the basis of the energy input (Fig. 10.13).

An NiMH battery can be used as energy storage as well as the supercapacitors. In Fig. 10.14 an example of a charging circuit. Two delay stages are used, controlled by U1 and U2 MOSFETs, to ensure sufficient stored energy from the boost converter before the gating in the battery. The power switches for U1 and U2 are Q1 and Q2, respectively. The diode D5 prevents any return trip of energy from the battery in the charging circuit when the charger is inactive.

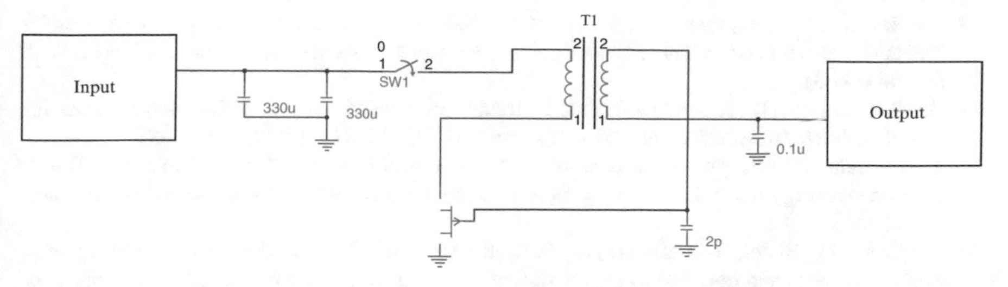

Fig. 10.13 Boost configuration

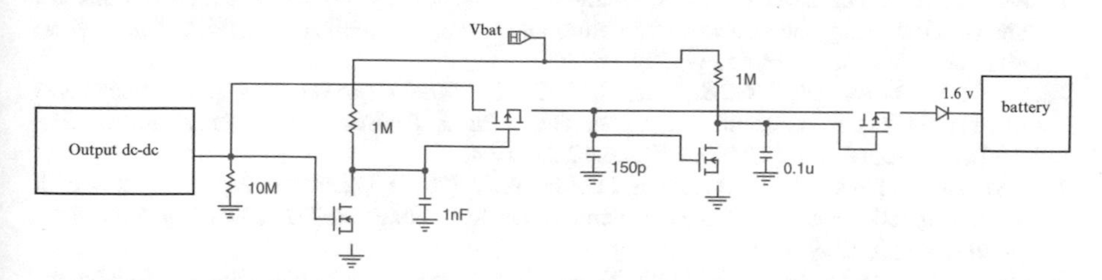

Fig. 10.14 Battery charging circuit

References

1. Park, J., & Mackqy, S. (2003). *Practical data acquisition for instrumentation and system control*. Oxford: Elsevier.
2. Lacanette, K. (2003). *National temperature sensors handbook*. Handbook Ann. Mat. National semiconductor.
3. National Instruments (1996). *Data Acquisition Fundamentals, Application Note 007*.
4. National Instruments (1996). *Signal Conditioning Fundamentals for PC-Based Data Acquisition Systems*.
5. Taylor, J. (1986). *Computer-based data acquisition system*. Research Triangle Park: Instrument Society of America.
6. Di Paolo Emilio, M. (2013). *Data acquisition system, from fundamentals to applied design*. New York: Springer.
7. Roundy, S., Wright, P., & Pister, K. (2002). Micro-electrostatic vibration-to- electricity converters. In *Proceedings of ASME International Mechanical Engineering Congress and Exposition IMECE2002* (Vol. 220, pp. 17–22).
8. Stordeur, M., & Stark, I. (1997). Low power thermoelectric generator: Self- sufficient energy supply for micro systems. In *Proceedings of the 16th International Conference on Thermo-Electrics* (pp. 575–577).
9. Shenck, N., & Paradiso, J. (2001). Energy scavenging with shoe-mounted piezoelectrics. *IEEE Micro, 21*(3), 30–42.
10. Roundy, S. (2003). *Energy scavenging for wireless sensor nodes with a focus on vibration to electricity conversion*. PhD thesis, University of California.
11. Tsutsumino, T., Suzuki, Y., Kasagi, N., Kashiwagi, K., & Morizawa, Y. (2006, November). Micro seismic electret generator for energy harvesting. In *Technical Digest PowerMEMS 2006*, Berkeley, USA (pp. 133–136).
12. Sterken, T., Altena, G., Fiorini, P., & Puers, R. (2007, April). *Characterisation of an electrostatic vibration harvester, DTIP of MEMS and MOEMS*, Stresa, Italy.

13. Sterken, T., Baert, K., Puers, R., & Borghs, S. (2002, November). Power extraction from ambient vibration. In *Proc. SeSens. Workshop on semiconductor sensors*, Veldhoven, Netherlands (pp. 680–683).
14. Szarka, G., Stark, B., & Burrow, S. (2012). Review of power management for energy harvesting systems. *IEEE Transactions on Power Systems, 27*(2), 803–815. ISSN: 0885-8993.
15. Cammarano, A., Burrow, S. G., Barton, D. A. W., Carrella, A., & Clare, L. R. (2010). Tuning a resonant energy harvester using a generalized electrical load. *Smart Materials and Structure, 19*, 055003.
16. Guyomar, D., Badel, A., Lefeuvre, E., & Richard, C. (2005). Toward energy harvesting using active materials and conversion improvement by nonlinear processing. *IEEE Transactions on Ultrasonics, Ferroelectrics, and Frequency Control, 52*, 584–595.
17. Mitcheson, P. D., Stoianov, I., & Yeatman, E. M. (2012). Power-extraction circuits for piezoelectric energy harvesters in miniature and low-power applications. *IEEE Transactions on Power Electronics, 27*, 4514–4529.
18. Szarka, G. D., Burrow, S. G., & Stark, B. H. (2012). Ultra-low power, fully-autonomous boost rectifier for electro-magnetic energy harvesters. *IEEE Transactions on Power Electronics, 28*(7), 3353–3362. doi:10.1109/TPEL.2012.2219594.
19. Maurath, D., Becker, P. F., Spreeman, D., Manoli, Y. (2012). Efficient energy harvesting with electromagnetic energy transducers using active low-voltage. *IEEE Journal of Solid-State Circuits, 47*(6), 1369–1380.
20. Beeby, S. P., Tudor, M. J., & White, N. M. (2006). Energy harvesting vibration sources for microsystems applications. *Measurement Science and Technology, 17*, R175–R195.
21. Khaligh, A., Zeng, P., & Zheng, C. (2010). Kinetic energy harvesting using piezoelectric and electromagnetic technologies–state of the art. *IEEE Transactions on Industrial Electronics, 57*(3), 850–860.
22. Paulo, J., & Gaspar, P. D. (2010). Review and future trend of energy harvesting methods for portable medical devices. In *Proceedings of the World Congress on Engineering* (Vol. 2).
23. Zhu, D., Tudor, M. J., & Beeby, S. P. (2010). Strategies for increasing the operating frequency range of vibration energy harvesters: A review. *Measurement Science and Technology, 21*, 022001-1–022001-29.
24. Cepnik, C., Lausecker, R., & Wallrabe, U. (2013). Review on electrodynamic energy harvesters – a classification approach. *Micromachines, 4*(2), 168–196. http://www.mdpi.com/2072-666X/4/2/168. Accessed 20 Jan 2015.
25. Ulaby, F. T., Michielssen, E., & Ravaioli, U. (2010). *Fundamentals of applied electromagnetics* (6th ed.). Upper Saddle River: Prentice Hall.
26. Roundy, S., Wright, P. K., & Rabaey, J. M. (2003). A study of low level vibrations as a power source for wireless sensor nodes. *Computer Communications, 26*(11), 1131–1144.
27. Sazonov, E., Li, H., Curry, D., & Pillay, P. (2009). Self-powered sensors for monitoring of highway bridges. *IEEE Sensors Journal, 9*, 1422–1429.
28. Toh, T. T., Mitcheson, P. D., Holmes, A. S., & Yeatman, E. M. (2008). A continuously rotating energy harvester with maximum power point tracking. *Journal of Micromechanics and Microengineering, 18*, 104008-1-7.
29. Howey, D. A., Bansal, A., & Holmes, A. S. (2011). Design and performance of a centimetre-scale shrouded wind turbine for energy harvesting. *Smart Materials and Structures, 20*, 085021.
30. Razavi, B. (2002). *Design of analog CMOS integrated circuits*. New York: McGraw-Hill.
31. Razavi, B. (2008). *Fundamentals of microelectronics*. Chichester: Wiley.
32. Sedra, A. S., & Smith, K. C. (2013). *Microelectronic circuits*. Oxford: Oxford University Press.
33. Gudan, K. (2015). *Ultra-low power 2.4GHz RF energy harvesting and storage system with 25dBm sensitivity*. San Diego: IEEE. ISBN 978-1-4799-1937-6.
34. Briand, D. (2015). *Micro energy harvesting*. Weinheim: Wiley.
35. Spies, P. (2015). *Handbook of energy harvesting power supplies and applications*. Boca Raton: CRC Press.

Chapter 11
Applications of Energy Harvesting

11.1 Introduction

Energy harvesting makes use of ambient energy to power small electronic devices such as wireless sensors, microcontrollers, and displays. Typical examples of these environmental sources are sunlight and any artificial source such as vibration or heat from engines or the human body. The energy transducers such as solar cells, thermogenerators, and piezoelectric convert this energy into electrical energy. The goal of each energy harvesting system is to replace the batteries used to power by extending the charging intervals for the storage element. A first field of application is the automation with self-powered switches. Further applications are the monitoring systems for large industrial plants or structural monitoring of huge buildings. Another promising market is the consumer area with purses, clothing which show the energy transducers integrated in the form of solar cells or TEG or RF transmitters to recharge consumer products such as mobile phones or audio players. A general overview of energy estimated to be harvested in different application is summarized in Tables 11.1 and 11.2, instead, compare the power consumption of different microcontrollers state of the art in some modes of operation.

11.2 Building Automation

The building automation systems are networks of electronic devices for the control of different functions. They manage lights, heaters, air conditioning systems, doors, valves, and safety systems. The benefits of building automation have reduced energy and maintenance costs, increased safety and comfort. Typical energy savings with the aid of building automation systems are about 30 %. A typical reduction is the wiring that may be reduced about 70 %. It also reduces operating costs, since no energy is required for devices powered by energy harvesting. In addition, it

M. Di Paolo Emilio, *Microelectronic Circuit Design for Energy Harvesting Systems*, DOI 10.1007/978-3-319-47587-5_11

Table 11.1 Power consumption for some devices

Applications	Power requirement
Standy	10 nW
RFID tag	10 uW
GSM	1 W
Bluetooth transceiver	10 mW
FW receiver	1 mW

Table 11.2 Activities of time and power consumption

Applications	Switch-on time (ms)	Power requirements (mW)
Radio	1	57
Microprocessor	Permanent	1
Energy management	Permanent	0.1
Radio + microprocessor, standby	Permanent	0.003
Mobile phone	Permanent	5

saves maintenance costs, because no batteries need to be replaced or recharged. Wireless transceivers or transmitters and low-power sensors are the main devices in the field of building automation devices powered from environmental sources. Typical sensors used are those of temperature, humidity, and pressure. In addition, light sensors, presence sensors, or motion are useful for the control of building functions. Other application devices are sensors for controlling the position of valves in the heating systems, ventilation, and air conditioning. The big advantage for all these sensors is the slow variation of the physical parameters to be monitored. Thus, the measurements and transmission of these data should only take place with a small duty cycle, leading to a low overall power consumption of sensors and transmitters. Wake-up times typical of such systems to make a reading of a sensor are 1, 10, or 100 s. Most of the time, these sensors are in sleep mode. Therefore, even a low energy consumption is obtained by reducing the amount of data sent or the number of transmissions. The goal of optimization is always a minimum of energy consumption to enable the energy harvesting with low cost hardware. The power of transmission is typical, for example, 10 mW, the frequency band more or less at 868 MHz. In building automation, light as well as the movement and the heat can be used to power electronic devices. In addition, the solar cells are used in buildings in combination with a supercapacitor or a rechargeable battery back-up for night operation. Sometimes, only a few hours of sunshine are sufficient to fully charge an energy storage element and ensure an operation of a specific device for many hours without illumination. As an additional safety feature, a signal can be transmitted to report the correct functioning of the system. The typical brightness in homes is between 100 and 500 lx. Most of the rules on health and safety work requiring a minimum illumination of 500 lx in the workplace. The minimum illumination time necessary to the day of the light will be about 30 % shorter with fluorescent light of the same brightness. The thermal gradients are a further option to use in building automation. Especially with radiators, air conditioning or hot water pipes. Due to the temperature around 20 °C, the heat sinks must be used to maintain a thermal gradient sufficient. When different temperature differences are present, the maximum power point trackers (MPPT) are useful to match the power management for the maximum power of the TEG. The vibrations can also be used in buildings to get electricity. The problem is that only a small vibration in small frequencies are available, which require large generators with large seismic masses. Typical 0.01 g and a frequency range between 1 and 10 Hz are values available [1–20].

11.3 Environmental Monitoring

The idea of monitoring is to measure physical parameters such as the temperature. The monitoring of the conditions can be used for all types of machines such as motors, pumps, fans, or compressors to determine any faults. It employs different types of sensors to measure operating parameters of the system. All these data must be collected and analyzed by microcontrollers. Some monitoring systems of the conditions provide an Ethernet connection to collect all the data and pass it to a central computer. Most feasible are wireless sensor networks, which transmit data by radio signals by eliminating the need of the installation of cables. The galvanic isolation is required in some applications, as well as, for example, in high power transmission systems. With the energy harvesting sensors, radio, actuators, processors, and displays is possible to power without batteries or at least without the need to recharge or replace them. Additionally, the installation in inaccessible areas, remote or dangerous, and integration into machines becomes possible. In addition to the transmission of measured values such as acceleration and temperature, directly displays the sensors can be applied to allow monitoring of the parameters of interest in the position of occurrence (Fig. 11.1).

11.4 Structural Health Monitoring

The structural health monitoring (SHM) is the discovery process of damage in components and construction. The goal of SHM is to improve the safety and reliability of aerospace, civil and mechanical infrastructure to predict and detect damage before they reach a critical state. Especially extreme events such as earthquakes and typhoons, heavy snow and storm or simply the aging of materials, and environmental degradation cause serious doubts on the integrity of any facilities. Structural monitoring (SHM) is the characterization process of existing structures to identify certain properties of the structure, in order to have information about the actual structure in which power to develop analytical models for assessment of the state of the structure or to assess changes in the structural behavior. The continued operation of civil structures and infrastructures such as bridges and buildings requires careful strategy to maintain an efficient structure and with a high level of

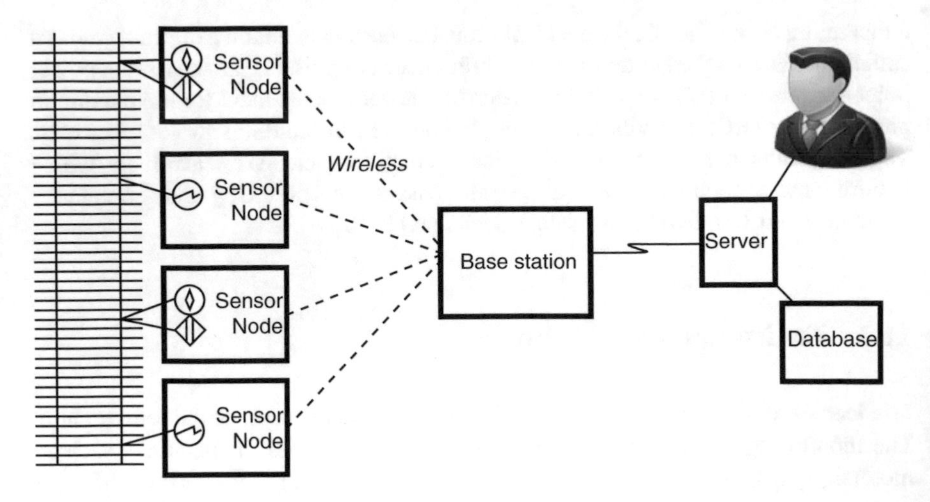

Fig. 11.1 WSN architecture

security but, at the same time, limiting maintenance costs. The problem becomes even more relevant when necessary in the evaluation of the performance level of potentially damaged by traumatic events structures, such as the seismic events. At the same time, safeguarding the immense architectural and cultural heritage represented by historic centers of many Italian cities, it requires careful monitoring of his state of health and the structural effects of the increasing level of vehicular traffic. The structures able to detect the operating temperatures, pressures which could reduce the weight and costs of composite materials. The weak point would connectors and cables for the passage of the sensor information to the outside of the structure. Wireless transmissions with extremely miniaturized electronics could solve this problem halfway, by leaving the question of power but with applicable techniques related to RF and thermal. The large number of individual measuring points and then large number of sensor nodes densely distributed in possibly random configurations in the environment detection require the use of wireless sensors. The cooperation between the sensor nodes is used as the local processing capability to perform the merge data or other calculation functions.

11.5 Automotive

The application of energy harvesting techniques in the railway sector is a very promising field. They were conducted studies and experiments on the potential of energy recovery devices, with the aim to provide information on the electric power actually generated by the use of harvester placed on board railway wagons. The tire pressure monitoring (TPM) is a means to increase the safety and efficiency of all types of vehicles by using tires inflated through monitoring the pressure.

Tire Pressure Monitoring systems were adopted in 1986 in a Porsche 959. The technology has been used in top luxury range of vehicles such as the Audi A6, Mercedes Benz S Class, and BMW7 Series to increase safety and the economy maintenance. Indirect TPMS do not use pressure sensors. These systems measure the air pressure in the wheel by monitoring individual wheel revolutions. Other developments may also simultaneously detect under-inflation until all four tires using vibration analysis of individual wheels or analysis of the effects of displacement of the load during acceleration or cornering. These systems require additional sensors suspension, which makes it more complex and expensive. To transfer the pressure and other data from the wheel to the vehicle control unit uses radio frequency (RF) communication or electromagnetic coupling. Due to the rotation of the wheel, no wired communication may be employed. The power of the sensor in the tire represents a challenge in direct TPM systems. Here batteries have the same problems in many other applications, such as limited life, low temperature range, weight, and additional costs. Due to the high amount of energy of vibration, TPMS are a potential application field for the vibration energy harvesting. The small required volume and low price facilitate the use of free self-powered systems. More suitable for use in tires are piezoelectric and electrostatic principles, mainly because of their small weight and size implementation related integration costs [21–30].

11.6 Projects

The system of Fig. 11.2 is designed to power a wireless sensor node that consumes 40 mW in active mode, with the average power of the load in according to the average power generated. The coupling of the temporal power is achieved by a supercapacitor and under-voltage lock-out which connects the load only when sufficient power is available for a cycle of operation. The topology of the power stage of the rectifier is shown in Fig. 11.3.

An advantage of the separation between start-up and main power circuits is that the charge pump can quickly establish a provision to support the requirements of the auxiliary circuitry, by allowing the rectifier to charge the super capacitor in an optimal way. The thermoelectric generators have recently been used as power supplies for strain gauges on planes. Strain gauges and the power supply are located outside the cab, and then are subjected to strong temperature variations in the course of a flight cycle. In order for the system to convert the greatest amount of latent heat energy to the possible change of phase in electrical form material, the thermoelectric generators are connected to a switching circuit capable of performing tracking the maximum power point (Fig. 11.4). A critical aspect is the inversion of the polarity between the ascent and descent of the aircraft, in this cases a rectifier is used in the system. It may not be a passive rectifier because the voltage required to effectively overcome the diode in the voltage state is not feasible for the TEG under these thermal gradients. Therefore, the rectifier must be one active with MOSFET.

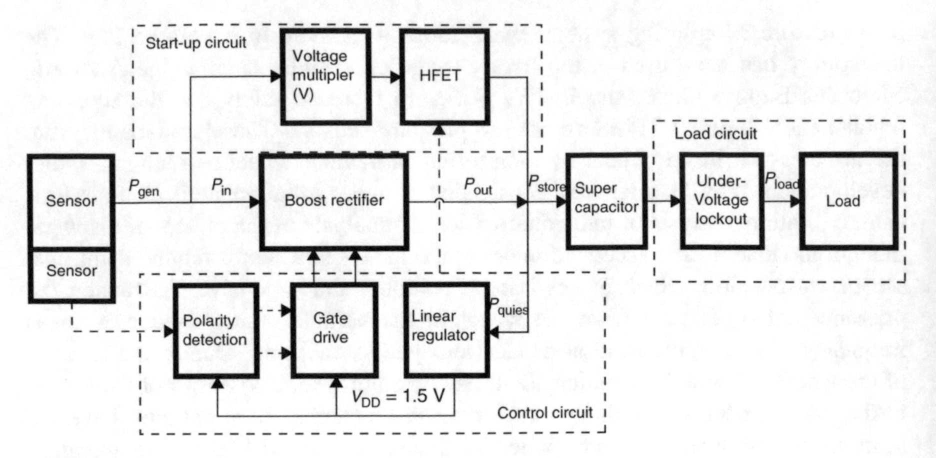

Fig. 11.2 Block diagram of an energy harvesting circuit

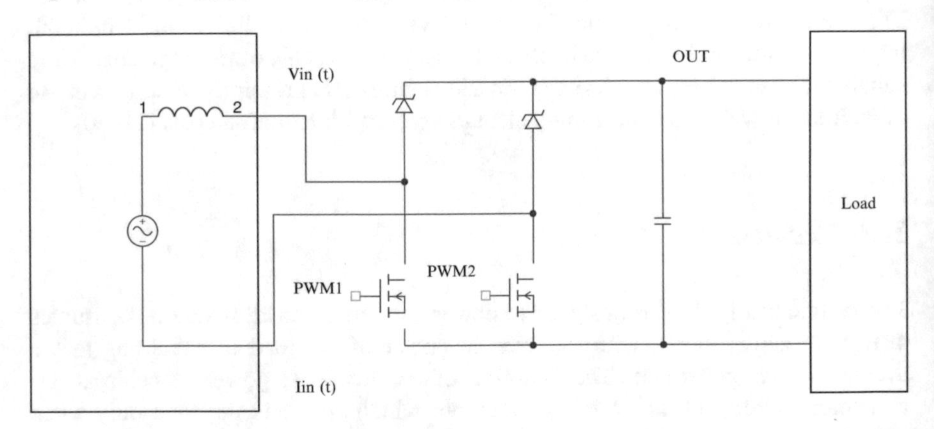

Fig. 11.3 Rectifier boost

11.7 Solar Infrastructure

Solar energy is derived from the electromagnetic waves produced by the sun. The solar radiation has the advantage to be available almost fully used without the production of polluting waste. The limits that may be encountered are discontinuities, caused by the alternation of day and night and weather conditions, the other is the low intensity which implies the need to have large areas of "energy storage." There are two different approaches for using solar energy. In the case of thermal utilization, the collectors transform the sun rays into heat energy that can be used both for producing hot water and to support the heating system. The sun's rays are converted into electrical energy using solar cells that are combined in modules within which a number of cells are connected in series and parallel combinations in order to obtain the voltage and the output current within the desired intervals. Also in this way

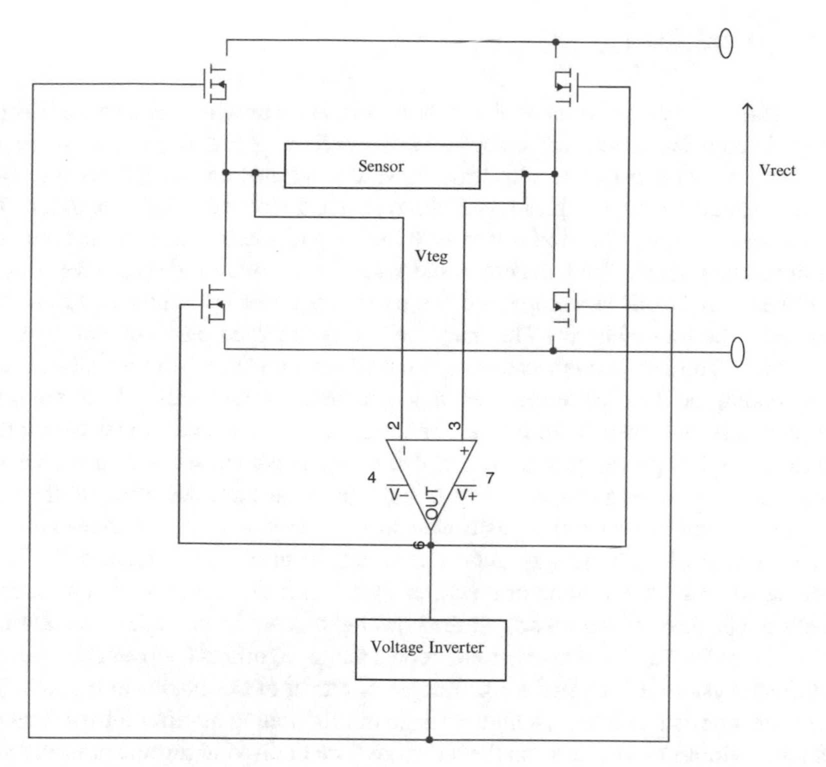

Fig. 11.4 TEG for airplane cabin

the power of a single module of practical utility reaches values (every single cell supplies only a few watts). The photovoltaic panels are systems that use lenses to concentrate the solar radiation on the cell, while still seeking to increase efficiency and reduce costs. A photovoltaic system is called "island" when it does not connect to any electrical service network and is therefore autonomous. Leverages on the place of production the energy collected and then preserved by an accumulator. The main components of a photovoltaic system "island" are generally PV array to gather energy through photovoltaic modules arranged in favor of the sun. By depending on the situation, the reference field is optimized to have a given voltage, usually either 12 or 24 V. Consequently, the majority of photovoltaic modules used in this type of plant has output voltages equal to 12 or 24 V, the so-called electrical strings that form the field are constituted of very few modules, up to the limit of the single module per string. Photovoltaic systems "grid-connect" are connected to an existing electricity distribution network and managed by third parties. Photovoltaic systems "hybrid" remain connected to the electrical distribution network, but mainly use solar energy and accumulator. Moreover, a control unit can be used by connecting the building to the electricity grid for the supply [31–35].

11.8 Wind Energy

The kinetic energy of wind is used to rotate a propeller or wheel, whose mechanical energy is then converted into electrical energy. Its exploitation, relatively simple and inexpensive, is implemented through wind machines divided into two distinct groups depending on the basic module type used defined wind generator. The kinetic wind energy utilization is very old: sailing and windmills constitute the most significant examples. The currently most widely used wind turbines have become integrated part of our landscape and the parts that make up a modern turbine are essentially the following: (1) The machine blades are fixed on a hub and make up in the rotor. The hub is connected to a rotating shaft that can be connected in turn, by depending on the type of turbines, to a synchronous generator or to a multiplier of revolutions and then to an asynchronous generator, currently most widespread solution. (2) The power cables. The above components are located in the nacelle positioned on a bearing support, so as to be easily adjustable according to the wind direction. L'entire spacecraft is positioned on a tower, which can be lattice or conical tubular. The most common system is to be varied the angle of incidence of the blade to the wind through hydraulic or electrical systems. A control system continuously checks the position of the blade and the power generated by changing continuously the angle of incidence. This method represents a significant advantage over the traditional system, which had a fixed angle of attack of the blades to the hub. The advantage consists in having a higher range of functioning at different wind speeds and lower vibrational stresses on the structure. The mini-wind turbine or even micro wind has the potential to convert the wind's kinetic energy into electrical energy, usable for the operation of electrical appliances throughout the home. In a particular way, for the installation of this type of wind power system, the low-power system from 0 to 20 kw, it can be installed in stand-alone systems to isolated users, in the best-known version of stand-alone, or in systems connected with the network, or the best known grid-connected.

11.9 Conclusions

The energy harvesting is adopted in many fields of applications, from consumer through automotive. Many of implementation scenarios differ with regard to the properties of the environmental energy source and the needs of electrical users. The economic benefits of harvesting became obvious when energy costs for battery replacement or recharging should be considered as the main factor in the design. Thus, energy harvesting is the preferred solution in applications where a single life of the battery of a non-rechargeable battery or a charge cycle is not sufficient. In applications with extreme conditions, it enables new solutions without the need to compete with batteries. Many quantities of energy harvesting devices are currently sold only in the building automation industry and in the consumer market. In the consumer market solar cells represent the main choice, but more solutions in terms of vibration and RF are entering the market. The rapidly expanding of wearable devices is set to revolutionize the technology and current processes in all areas, by creating new market opportunities and new business models. The market will continue to evolve as several start-ups and established companies have a strong interest in research and development, in particular for implementing multiple functions on a single device. The development of IoT promises to be exciting and have a great economic impact due to the development of semiconductor devices, and to the advancement of wireless technology, by allowing smaller devices and more efficient at the same time. IoT is a fundamental requirement for power management: mobile devices obviously require batteries, but the possibility of replacing them completely or restrict the replacement/charging is a factor of considerable importance given the advent of further devices connected in the near future. Thanks to new small autonomous devices that can retrieve small amounts of energy arising from vibrations from the motorcycle and the electromagnetic waves in the ether (such as those of mobile phones or TV). Several prototypes have been designed from shoes with electromagnetic recovery to tensile structures in the road coverage that can capture the energy derived from the traffic drafts; even the clothing capable of incorporating the energy produced by the movement of legs and arms, up to road surfaces able to absorb the vibrations of the vehicles and transform them into electrical energy. Mechanical vibrations, which are continually subject the streets and sidewalks, generate differences in electrical potential in piezoelectric devices "drowned" in areas subject to the passage of vehicles or pedestrians. The goal is to suppy in a sustainable way the lights sensors for traffic monitoring, street lighting, illuminated signs. A futuristic scenario is from one side to do devices without batteries, in many research centers are doing studies to optimize the power consumption in processors and electronic devices, and on the other side to recover the energy already available in the surrounding environment to be able to store and reuse for the self-supply in energy-efficient technologies such as wireless networks, sensors, LCD screens, MP3 players, GPS systems, and mobile phones.

References

1. Park, J., & Mackqy, S. (2003). *Practical data acquisition for instrumentation and system control*. Oxford: Elsevier.
2. Lacanette, K. (2003). *National temperature sensors handbook*. Handbook Ann. Mat. National semiconcductor
3. National Instruments (1996). Data acquisition fundamentals, Application Note 007.
4. National Instruments (1996). *Signal conditioning fundamentals for PC-based data acquisition systems, Handbook National Instruments*.
5. Taylor, J. (1986). *Computer-based data acquisition system*. Research Triangle Park: Instrument Society of America.
6. Di Paolo Emilio, M. (2013). *Data acquisition system, from fundamentals to applied design*. New York: Springer.

7. Roundy, S., Wright, P., & Pister, K. (2002). Micro-electrostatic vibration-to- electricity converters. In *Proceedings of ASME International Mechanical Engineering Congress and Exposition IMECE2002* (Vol. 220, pp. 17–22).
8. Stordeur, M., & Stark, I. (1997). Low power thermoelectric generator: Self- sufficient energy supply for micro systems. In *Proceedings of the 16th International Conference on Thermo-Electrics* (pp. 575–577).
9. Shenck, N., & Paradiso, J. (2001). Energy scavenging with shoe-mounted piezoelectrics. *IEEE Micro, 21*(3), 30–42.
10. Roundy, S. (2003). *Energy scavenging for wireless sensor nodes with a focus on vibration to electricity conversion*. PhD thesis, University of California.
11. Tsutsumino, T., Suzuki, Y., Kasagi, N., Kashiwagi, K., & Morizawa, Y. (2006, November). Micro seismic electret generator for energy harvesting. In *Technical Digest PowerMEMS 2006*, Berkeley, USA (pp. 133–136).
12. Sterken, T., Altena, G., Fiorini, P., & Puers, R. (2007, April). *Characterisation of an electrostatic vibration harvester, DTIP of MEMS and MOEMS*, Stresa, Italy.
13. Sterken, T., Baert, K., Puers, R., & Borghs, S. (2002, November). Power extraction from ambient vibration. In *Proc. SeSens. Workshop on semiconductor sensors*, Veldhoven, Netherlands (pp. 680–683).
14. Szarka, G., Stark, B., & Burrow, S. (2012). Review of power management for energy harvesting systems. *IEEE Transactions on Power Systems, 27*(2), 803–815. ISSN: 0885-8993.
15. Cammarano, A., Burrow, S. G., Barton, D. A. W., Carrella, A., & Clare, L. R. (2010). Tuning a resonant energy harvester using a generalized electrical load. *Smart Materials and Structure, 19*, 055003.
16. Guyomar, D., Badel, A., Lefeuvre, E., & Richard, C. (2005). Toward energy harvesting using active materials and conversion improvement by nonlinear processing. *IEEE Transactions on Ultrasonics, Ferroelectrics, and Frequency Control, 52*, 584–595.
17. Mitcheson, P. D., Stoianov, I., & Yeatman, E. M. (2012). Power-extraction circuits for piezoelectric energy harvesters in miniature and low-power applications. *IEEE Transactions on Power Electronics, 27*, 4514–4529.
18. Szarka, G. D., Burrow, S. G., & Stark, B. H. (2012). Ultra-low power, fully-autonomous boost rectifier for electro-magnetic energy harvesters. *IEEE Transactions on Power Electronics, 28*(7), 3353–3362. doi:10.1109/TPEL.2012.2219594.
19. Maurath, D., Becker, P. F., Spreeman, D., Manoli, Y. (2012). Efficient energy harvesting with electromagnetic energy transducers using active low-voltage. *IEEE Journal of Solid-State Circuits, 47*(6), 1369–1380.
20. Beeby, S. P., Tudor, M. J., & White, N. M. (2006). Energy harvesting vibration sources for microsystems applications. *Measurement Science and Technology, 17*, R175–R195.
21. Khaligh, A., Zeng, P., & Zheng, C. (2010). Kinetic energy harvesting using piezoelectric and electromagnetic technologies–state of the art. *IEEE Transactions on Industrial Electronics, 57*(3), 850–860.
22. Paulo, J., & Gaspar, P. D. (2010). Review and future trend of energy harvesting methods for portable medical devices. In *Proceedings of the World Congress on Engineering* (Vol. 2).
23. Zhu, D., Tudor, M. J., & Beeby, S. P. (2010). Strategies for increasing the operating frequency range of vibration energy harvesters: A review. *Measurement Science and Technology, 21*, 022001-1–022001-29.
24. Cepnik, C., Lausecker, R., & Wallrabe, U. (2013). Review on electrodynamic energy harvesters – a classification approach. *Micromachines, 4*(2), 168–196. http://www.mdpi.com/2072-666X/4/2/168. Accessed 20 Jan 2015.
25. Ulaby, F. T., Michielssen, E., & Ravaioli, U. (2010). *Fundamentals of applied electromagnetics* (6th ed.). Upper Saddle River: Prentice Hall.
26. Roundy, S., Wright, P. K., & Rabaey, J. M. (2003). A study of low level vibrations as a power source for wireless sensor nodes. *Computer Communications, 26*(11), 1131–1144.
27. Sazonov, E., Li, H., Curry, D., & Pillay, P. (2009). Self-powered sensors for monitoring of highway bridges. *IEEE Sensors Journal, 9*, 1422–1429.

28. Toh, T. T., Mitcheson, P. D., Holmes, A. S., & Yeatman, E. M. (2008). A continuously rotating energy harvester with maximum power point tracking. *Journal of Micromechanics and Microengineering, 18*, 104008-1-7.
29. Howey, D. A., Bansal, A., & Holmes, A. S. (2011). Design and performance of a centimetre-scale shrouded wind turbine for energy harvesting. *Smart Materials and Structures, 20*, 085021.
30. Razavi, B. (2002). *Design of analog CMOS integrated circuits*. New York: McGraw-Hill.
31. Razavi, B. (2008). *Fundamentals of microelectronics*. Chichester: Wiley.
32. Sedra, A. S., & Smith, K. C. (2013). *Microelectronic circuits*. Oxford: Oxford University Press.
33. Gudan, K. (2015). *Ultra-low power 2.4GHz RF energy harvesting and storage system with 25dBm sensitivity*. San Diego: IEEE. ISBN 978-1-4799-1937-6.
34. Briand, D. (2015). *Micro energy harvesting*. Weinheim: Wiley.
35. Spies, P. (2015). *Handbook of energy harvesting power supplies and applications*. Boca Raton: CRC Press.

Index

A
AC-DC, 76
active sensors, 2
Amplifier, 131
amplifier, 120
Antenna, 41, 144
Applications, 155
Armstrong oscillator, 94
Atoms, 37
Automotive, 158

B
BAN, 146
Bandwidth, 132
Battery, 143
BiCMOS, 145
BJT, 111
Boltzmann, 57
Boost converter, 91
Bridge rectifier, 79
Buck converter, 91
Buffer, 98
Building automation, 155

C
Capacitor, 66
Carnot, 58
Channel, 129
Circular spiral antenna, 148
CMOS Inverter, 125
Comb drive, 70
Common drain MOS, 131
Common gate MOS, 131, 132
Common source MOS, 131
Conditioning, 76
Conductors, 55
Constan-charge, 69
Continuos system, 67
Converter, 76
Cosmic rays, 27
Current Mirror, 125

D
DAQ, 1
DC-DC, 41, 76
Diffusion, 57
digital, 1
Diode, 30, 42, 110
Donors, 57

E
Earth, 28
Efficiency, 60
electromagnetic wave, 37
electrostatic transducer, 65
Emitter follower, 113
energy, 1, 11
Entropy, 58
Enviromental monitoring, 157
Ethernet, 157
Excitation piezo, 51

F
Feedback, 122
Ferrite, 146

M. Di Paolo Emilio, *Microelectronic Circuit Design for Energy Harvesting Systems*, DOI 10.1007/978-3-319-47587-5

FET, 129
field wiring, 7
Figure of merit, 61
flicke noise, 7
Folded dipole, 149
Forward Bias, 110
Friss, 37
Full-wave circuit, 78

G
Gap closing, 70
Gradient, 57

H
Half-wave circuit, 78
HVAC, 15

I
inertial motion, 65
IoT, 162

J
Junctions, 55

L
L network, 41
langasite, 48
light sensors, 5
Linear regulators, 89
LNA, 139
load, 13
Load matching, 95
Low noise pre-amplifier, 139

M
magnetic field, 4
magnetic sensors, 4
Matching circuit, 42
Maxwell, 37
Mechanical, 65
mechanical stress, 48
memory board, 3
MEMS, 22, 66
micro-generators, 66
Microelectronic, 105
Micropowering, 78
Microstrip, 149
Miller, 132
Mobile, 26, 146
Mosfet, 69, 116, 129
MPP, 32
MPPT, 87
Multi-layer, 61

N
NMOS, 124
nMosfet, 129
noise, 7
Nyquisit, 124

O
Operational amplifier, 134
Output response, 132

P
Parallel plate, 70
passive sensors, 2
Peacemakers, 144
Peltier, 55
Photovoltaic, 160
Photovoltaic cell, 30
piezoceramic, 47
Piezoelectric, 22, 69
piezoelectric, 47
Piezoelectric frequency, 49
piezoelectric model, 48
Plank, 30
PMOS, 124
pMosfet, 129
polycrystalline, 48
potentiometer, 5
Power management, 150
power management, 15
PSRR, 15, 136
PZT, 47

Q
Q factor, 42

R
Rectenna, 146
Rectifier, 42, 76
rectifier, 13
Rectifier circuits, 78
Reverse Bias, 109
RF harvesting, 151
Rolling rod, 70
RTD, 3

S

Schottky, 42
Seebeck, 24, 55
semiconductor, 106
Semicondutors, 57
sensors, 1
settling time, 8
SHM, 157
shot noise, 7
Silicon, 69
silicon, 106
Silicon dioxide, 129
SNR, 120
Solar, 88
Solar infrastructure, 160
Solar radiation, 27
Solar wind, 27
Space radiation, 27
storage, 18
Sun spectrum, 28
Supercapacitor, 26
Supercapacitors, 98
Switched system, 67
Switching regulators, 90

T

Teflon, 69
TEG, 143, 155
Thermal, 22
thermal, 12
Thermal gradient, 24
Thermal noise, 7
thermistors, 3
Thermocouple, 59
thermocouples, 3
Thermodynamics, 59
Thermoelectric, 55
Thermoelectric generator, 59
Thermogenerators, 155
thin film, 105
TPM, 158
transconductor, 132
transducers, 1

V

Variable overlap, 70
Vibration, 21
vibration, 11
Vibrations, 49
Voltage control, 86
voltage regulator, 13

W

Wearable, 143
Wi, 37
Wind infrastructure, 160
Wireless sensors, 157
WSN, 18

Y

Young, 49

Z

Zener diode, 81

Printed in the United States
By Bookmasters